Foreword

Inside this book is a U.S. military-designed crash course in woodworking. While it definitely wasn't written for the hobbyist in 1946, the information presented here is extraordinarily helpful to anyone looking to get started in woodworking. From the basics of how wood works, to an overview of different types of joints used in making furniture, to plans and material lists for 28 different furniture projects, this book is packed with useful information for both the novice and advanced woodworker.

While some of the tools here don't look like their modern-day counterparts, surprisingly little has changed in the world of woodworking since the end of WWII. Table saws are still flat surfaces with spinning blades and fences, lathes still spin wood so it can be shaped with chisels and gouges, and wood moves in the same ways. It's true that radial arm saws aren't as prevalent as they were back then. And there are (thankfully) more safety features in modern equipment.

The furniture pieces depicted in the plans are an interesting combination of basic, relatively unadorned forms with a few Colonial Revival touches. While you may not want to replicate these pieces precisely, you'll get insight into how casework, chairs and tables should be constructed to hold up to the rigors of life on a military base.

Taken as a historic text, it's entertaining to see black-and-white images of men with fedoras and ties (safely tucked out of harm's way) demonstrating proper, state-sanctioned woodworking techniques. I can just imagine being a new recruit tasked with caring for and maintaining the furniture on base, with this manual guiding my way. Whether you're a new recruit to the world of woodworking, or you're a seasoned soldier, I hope you'll enjoy this look back as much as I do.

Andrew Zoellner
Editor in Chief
Popular Woodworking

Distributed in the U.K. and Europe by
F&W Media International
Pynes Hill Court
Pynes Hill
Rydon Lane
Exeter
EX2 5AZ
United Kingdom
Tel: (+44) 1392 797680

Visit our website at popularwoodworking.com for more woodworking information.

Other fine Popular Woodworking Books are available
from your local bookstore or direct from the publisher.

ISBN-13: 978-1-4403-5506-6

23   22   21   20   19   5   4   3   2   1

EDITOR: David Thiel
COVER DESIGNER: Danielle Lowery
PRODUCTION MANAGER: Debbie Thomas

fw
a content + ecommerce company

CONTENTS

			Paragraphs	*Page*
CHAPTER	1.	INTRODUCTION	1-2	1
CHAPTER	2.	WOOD	3-8	2
CHAPTER	3.	FURNITURE CONSTRUCTION	9-10	10
CHAPTER	4.	WOODWORKING MACHINERY		54
		Section I. INTRODUCTION	11	54
		II. CIRCULAR SAW BENCHES	12-22	54
		III. OVERARM SAW	23-35	75
		IV. WOOD JOINTER	36-40	84
		V. SINGLE SURFACER	41-43	88
		VI. WOOD SHAPER	44-46	93
		VII. BAND SAW	47-48	96
		VIII. HOLLOW-CHISEL MORTISER	49-50	100
		IX. WOOD LATHE	51-53	103
CHAPTER	5.	GLUES AND GLUING TECHNIQUE	54-61	108
CHAPTER	6.	WOOD FURNITURE	62-68	114
CHAPTER	7.	UPHOLSTERED FURNITURE	69-74	117
CHAPTER	8.	REFINISHING	75-80	120
CHAPTER	9.	METAL FURNITURE	81	125
APPENDIX	I.	DEFINITIONS		126
	II.	REFERENCES		127
INDEX				128

WAR DEPARTMENT
WASHINGTON 25, D. C., 1 June 1946

TM 5-613, Woodworking and Furniture Repair, Repairs and Utilities, is published for the information and guidance of all concerned.

[AG 300.7 (13 Mar. 46).]

BY ORDER OF THE SECRETARY OF WAR:

OFFICIAL:
EDWARD F. WITSELL
Major General
The Adjutant General

DWIGHT D. EISENHOWER
Chief of Staff

DISTRIBUTION:

AAF (2); AGF (2); ASF (2); T (Att Eng) (10); Dept (Att Eng (5); Base Comd (10); Def Comd (Att Eng) (10); AAF Comd (10); HD (Att Eng) (5); S Div ASF (1); Tech Sv (2) except OCE (75); SvC (Att Eng) (10); FC (Att Post Eng) (5); Class III Instls (Att Post Eng) (5); PE (Att Eng) (5); Sub-PE (Att Eng) (5); Ars (Att Post Eng) (5); Dep (Att Eng) (5); Dist 5 (OCE) (2); Div Eng (OCE) (10); GH (Att Post Eng) (5); RH (Att Post Eng) (5); CH (Att Post Eng) (5); Tng C (ASF) (5); Rehab C (Att Eng) (5); PW Cp (Att Post Eng) (5); D (Eng) (2); Bn (Overseas only) (Eng) (2)

Refer to FM 21-6 for explanation of distribution formula.

CHAPTER I

INTRODUCTION

1. Purpose and Scope

This Technical Manual is a guide to repair of furniture used in offices, clubs, messes, quarters, and hospitals on Army posts. It covers construction of the more common pieces of furniture, and describes tools and techniques used in furniture repair. It also includes a brief discussion of the properties and uses of various types of wood. (For policy governing furniture repair see TM 5-600.)

2. Standards

The following standards govern all repair work:

a. MATERIALS. Use materials similar in quality to those used in the original article.

(1) *Wood.* Use properly seasoned wood. (See par. 5.) Match the species, color, grain, and texture of other members as closely as possible. Make sure the grain of the wood is approximately continuous across any lapped or spliced joint; if the grain in the member to be repaired cannot be matched, replace the entire member.

(2) *Fabrics.* Match closely the pattern, color, and texture of fabric in the original piece.

b. WORKMANSHIP. Use good workmanship, regardless of the nature of the repair. See that joints are strong and that surfaces are in condition to receive an acceptable finish. Give furniture a finish suited to its intended use.

CHAPTER 2

WOOD

3. General

A general understanding of tree classification and wood structure helps the woodworker make the best use of his materials.

a. CLASSIFICATION OF TREES. Nearly all trees are included in the most important of the four major plant groups, the Spermatophyta. This group is divided into Gymnosperms and Angiosperms.

(1) The Gymnosperms include the conifers, known in the lumber industry as softwoods.

(2) The Angiosperms include the monocotyledons (palms, yuccas, bamboos) and the dicotyledons. The latter are much more important. They include all hardwood trees.

(3) The terms hardwood and softwood should be used only in considering a group as a whole. Actually, the wood of certain softwood trees is harder than that of certain hardwood trees.

b. TREE STRUCTURE. A tree is a complex structure of roots, trunk, limbs, and leaves. Only the larger portion of the trunk or bole is used for lumber. This portion is first crosscut into logs. Figures 1 and 2 illustrate the important parts of wood.

(1) *Tissue zones.* A cross section of a tree trunk shows the following well-defined tissue zones in succession from the outside to the center: bark, wood, and pith, a small spot at the center, usually darker in color than the wood. (See fig. 1.)

(2) *Heartwood and sapwood.* In most species, wood at the center of the trunk (heartwood) is darker than wood in the outer part (sapwood) and varies from it slightly in physical properties. The relative proportions of heartwood and sapwood in a tree vary with species and environment. Sapwood normally can be seasoned more easily than heartwood. It is more susceptible to fungus and insect attacks, but is more easily impregnated with wood preservatives. There is no difference in strength.

c. TREE GROWTH. Wood is not a homogeneous substance, but a composite structure similar in many respects to a honeycomb. It is formed by the accretion of countless numbers of small units known as cells.

(1) *Formation of growth rings.* When growth is interrupted each year by cold weather or drought, the structure of cells formed at the end and at the beginning of the growing season is different enough to define sharply the annual layers or growth rings. (See fig. 2.)

(2) *Structure of growth rings.* In many species, each annual ring is divided more or less distinctly into two layers. The inner one, the springwood, consists of cells having relatively large cavities and thin walls. The outer layer, the summerwood, is composed of smaller cells. The transition from springwood to summerwood may be abrupt or gradual, depending on the kind of wood and growing conditions at the time it was formed. In most species, springwood differs from summerwood in physical properties, being lighter in weight, softer, and weaker. Species such as the maples, gums, and poplars do not show much difference in the structure and properties of the wood formed early or later in the season.

d. STRENGTH. Strength of wood depends on the species, growth rate, specific gravity, and moisture content. Extremely slow growth produces a weaker wood. Softwoods (conifers) also are weakened by extremely rapid growth. Wood with low specific gravity or high moisture content is generally weaker. Defects such as grain deviation caused by spiral growth, knots, and burls, also result in weaker wood.

e. APPEARANCE. Structural defects frequently enhance the appearance of wood. Spiral growth results in a winding stripe on turnings. Butt wood shows the assembly of root branches and crotch wood has a merging or diverging pattern. A burl produces attractive boards showing tissue distortion. The bird's-eye figures resulting from the elliptical arrangement

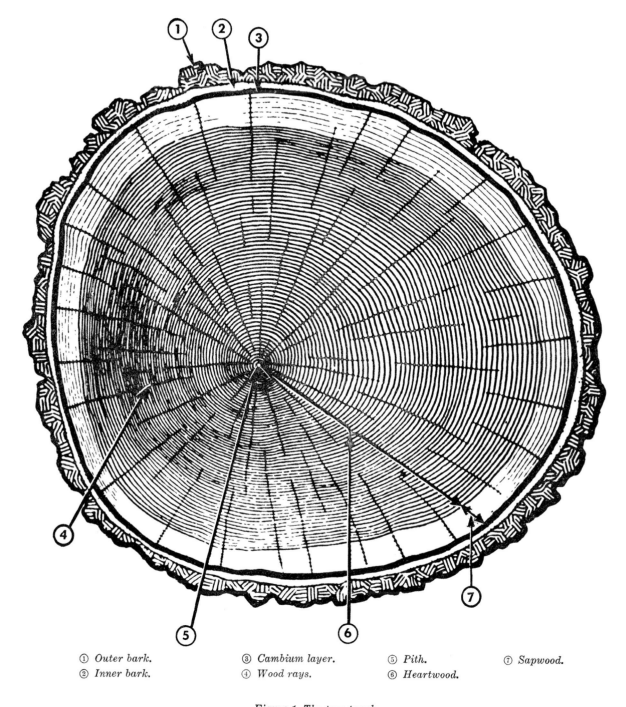

① Outer bark.   ③ Cambium layer.   ⑤ Pith.   ⑦ Sapwood.
② Inner bark.   ④ Wood rays.   ⑥ Heartwood.

Figure 1. The tree trunk.

of wood fibers around a series of central spots do not weaken maplewood appreciably. Some quarter-sawed woods show pronounced whitish flakes where the wood rays are exposed. This forms an interesting pattern, especially in oak and sycamore.

f. COMPOSITION OF WOOD. Wood is composed mainly of cellulose and lignin. Cellulose is the basic substance; the lignin acts as a stiffening and bonding agent.

Cross section of a log, showing annual growth rings.   *Light rings are springwood; dark rings, summerwood.*

Figure 2.

4. Lumber

a. WAYS OF CUTTING. Lumber is sawed from a log in two distinct ways, with the plane of the cut either radial or tangential to the annual rings.

(1) When the cut is tangent to the annual rings, the lumber is known as plain-sawed (hardwoods) or flat-grain lumber (softwoods). (See fig. 3.)

(2) When the cut is in a radial plane (parallel to the wood rays) the lumber is known as quarter-sawed lumber (hardwoods), or edge or vertical grain lumber (softwoods). (See fig. 3.)

(3) It is commercial practice to call lumber with annual rings at angles from 45° to 90° with the board surface "quarter-sawed," while lumber with rings at angles from 0° to 45°

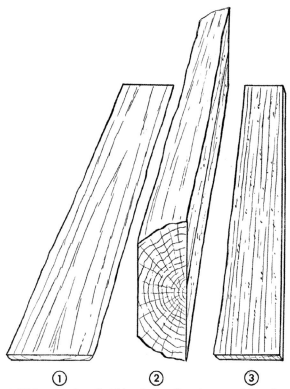

1. Plain-sawed. 2. Cut from log. 3. Quarter-sawed boards.

Figure 3.

with the surface is called "plain-sawed." Unless the logs are very large, the average sawmill output consists mainly of plain-sawed lumber.

b. RELATIVE ADVANTAGES. The relative advantages of plain-sawed and quarter-sawed lumber are:

(1) *Plain-sawed lumber.* (a) Usually cheaper than quarter-sawed lumber because it can be cut from the log faster and with less waste.

(b) Is less likely to collapse in drying.

(c) Shakes and pitch pockets extend through fewer boards.

(d) Round or oval knots affect surface appearance and strength less than the spike knots in quarter-sawed boards.

(e) Figures formed by annual rings and other grain deviations are more conspicuous.

(2) *Quarter-sawed lumber.* (a) Shrinks and swells less in width than plain-sawed lumber.

(b) Cups and twists less.

(c) Does not surface-check or split as badly in seasoning and use.

(d) Wears more evenly.

(e) Raised grain caused by the annual rings is not as pronounced.

(f) Is less pervious to liquids.

(g) Most species hold paint better.

(h) Figures resulting from pronounced rays, interlocked grain, and wavy grain are more conspicuous.

(i) Width of the sapwood appearing in a board is no greater than that of the sapwood ring in the log.

5. Seasoning Lumber

a. PURPOSE OF SEASONING. As it comes from the sawmill, lumber has a high moisture content and is unsuited for most shop use. Moisture content is the weight of water contained in the wood, expressed as a percentage of the weight of the oven-dry wood. It is important that lumber be seasoned until the moisture content is in equilibrium with the conditions under which the wood will be in service. When a condition of equilibrium moisture content is reached, the lumber has no tendency to shrink, expand, or warp. Because of normal changes in atmospheric moisture, this condition never holds constant. It is desirable, however, that an approximate equilibrium moisture content be reached.

(1) *Causes of shrinkage and expansion.* As the moisture content of a piece of wood decreases, the wood shrinks. Shrinkage begins when the moisture content drops below the fiber-saturation point. In most woods this is about 30 per cent, the moisture content at which all free water disappears from the cell cavity, while the cell walls are still saturated. Normal air-seasoning practices reduce the moisture content of lumber to between 12 and 15 per cent. If this wood is made into an article which is subjected to further drying action, shrinkage continues. Conversely, if the moisture content of a piece of wood is increased, the piece swells.

(2) *Extent of shrinkage and expansion.* Wood expands or shrinks only 0.01 to 0.02 of 1 per cent along the grain in length, but can change considerably across gain in width and

5

thickness. Limits vary by species between 4 and 14 per cent for tangential shrinkage and between 2 and 8 per cent for radial shrinkage. The variation in expansion is about the same. Even when painted, wood continues to absorb and lose moisture according to long-time changes in atmospheric humidity. Therefore, it is always good policy to keep a small stock of lumber in the room in which it is going to be worked.

b. DETERMINATION OF METHOD. Lumber can be seasoned by natural air-drying, by kiln-drying, or by various chemicals (common salt, urea, and so on) in combination with the first two methods. The time available for drying, the species of wood, and the ultimate use of the wood are important factors in determining the method of seasoning.

(1) *Time.* If the lumber must be dried and ready for use in a limited time, it is seasoned by kiln-drying. Depending on species and size of stock, kiln-drying requires at least 3 days, while air seasoning normally requires at least 2 months.

(2) *Species.* Some species need special treatment to prevent checking and warping. For example, black and tupelo gum boards can be kiln-dried slowly with excellent results, but usually warp badly when air-dried if the lumber is not properly piled.

(3) *Ultimate use.* The moisture content of cabinet woods must be within a range of 5 to 7 per cent. Since this percentage cannot be reached with normal air seasoning, such wood must be kiln-dried.

c. KILN-DRYING. Kiln-drying is the common commercial practice.

(1) *Types of kilns.* The modern compartment kiln has positive control over temperature, relative humidity, and air movement. Progressive type kilns are satisfactory for some kinds of lumber but are not so readily controlled as compartment kilns.

(2) *Effects of improper kiln-drying.* Seasoning defects are made worse by improper kiln-drying. The most common seasoning defects caused by faulty kilns or poor operation are:

(*a*) Casehardening, a condition in which such stresses are developed in a board that the outside is in compression and the inside is in tension even when both are equally dry. A flat casehardened board cups when sawed.

(*b*) Honeycombing (interior checking).

(*c*) Collapse of cell structure, manifested by an irregular cross section and surface corrugations.

(*d*) Warping in any or all of four forms:

1. Crook—edge of board convex or concave along its length.

2. Bow—face of board convex or concave along its length.

3. Cup—face of board convex or concave across its width.

4. Twist—the turning or winding of the edge of a board so that the four corners of any face are no longer in the same plane.

(*e*) Brashness, a condition characterized by low resistance to shock; caused in part by exposure to excessive heat.

(*f*) Surface and end checking (longitudinal radial cracks or separations formed to relieve drying stresses).

d. CHEMICAL SEASONING. Chemical seasoning in combination with air- or kiln-drying is used by certain lumber companies for seasoning high-quality lumber. Green lumber is taken directly from the grading table and treated with one of several chemicals. Urea is relatively cheap and has proved particularly effective in kiln-drying Douglas fir timbers. There should be a fairly high concentration of the chemical in the surface of the lumber. The zone of chemical concentration has a lower vapor pressure than water, so the surface layer is water-hungry and pulls the moisture from the interior of the board to the surface. Theoretically, the surface zone remains moist at the expense of the interior. This is a reversal of the usual method of seasoning, in which the surface dries first. It sometimes makes possible seasoning of large timbers without the serious surface checking formerly accepted as inevitable.

6. Defects in Lumber

Most lumber has some defect. The most common defects are:

a. WARP. (See par. 5*c*.)

b. GRAIN DEVIATION. Grain deviation is a condition in which the grain does not run parallel to the plane of the board. It may be due to natural or artificial causes. Spiral, wavy, curly, or bird's-eye figures and distortions caused by an injury or knot are natural grain deviations. Artificial deviations result when the plane of the saw cut is not parallel to the outside surface of the log. Grain deviation weakens a board; so, if strength is an important consideration, lumber with grain deviations must not be used, particularly if the grain slope exceeds 1 in 15.

c. KNOTS. Knots are portions of what once were limbs. As the tree grows, they become embedded in the trunk. If the limb is alive when the tree is cut, the knot is sound and tight. If the limb is dead, the knot is usually loose, at least in the part of the trunk that grew after the limb died. The grain deviation around a knot, rather than the knot itself, weakens lumber. Knots are round, oval, or spike-shaped, depending on whether the log is plain- or quarter-sawed.

d. COMPRESSION WOOD. Compression wood is frequently found in conifers on the underside of branches and leaning trunks. It is often characterized by an eccentric pattern of annual rings with an excess of opaque summerwood. The higher specific gravity of compression wood does not mean that this wood is stronger than normal wood. On the contrary, it is weaker and shrinks more in length.

e. INJURIES FROM HANDLING. If a log is dropped across a rock, or a falling tree strikes a stump, the fibers are likely to be crushed and much of their tensile strength is lost. This defect, known as compression failure, appears on a board as a fine irregular line across the grain.

f. SHAKES. A shake is a separation of wood along the annual ring. All causes of this defect are not known.

g. CHECKS. See paragraph 5c.

h. MOLDS, STAINS, WOOD ROTS. Wood is subject to the destructive action of a large number of fungi. These are microscopic saprophytic plant growths. Molds attack only the surface of the wood, and do little more damage than a layer of dirt. Since penetration is slight, molds often may be dressed off lumber. Stain fungi discolor wood but do not destroy much of the structure. Wood-rotting fungi break down the wood structure and in time reduce the wood to dust. Warmth, moisture, and air are necessary for fungus growth.

i. INSECT DAMAGE. Insect damage to seasoned or even partially seasoned lumber is usually slight. Certain woods are susceptible to Lyctus (powder-post) beetles. If a powdery substance is noticed coming from small holes, these beetles have attacked the wood. (See TM 5-632.)

7. Preservation

This term usually refers to the treatment of wood to make it resistant to fungi and insects. The most common method is to treat the wood with a substance poisonous to the destructive organism. Creosote, salts such as zinc chloride, and organic compounds such as beta naphthol are effective.

8. Furniture Woods

The following characteristics are desirable in wood used to make or repair furniture:

a. Stability, or ability to keep its shape without shrinking, swelling, or warping.

b. East of fabricating, surfacing, and finishing.

c. Pleasing appearance.

d. Suitable strength and grain characteristics.

e. Availability.

Table I lists properties and uses of common furniture woods.

TABLE I. *Properties and uses of common furniture woods*

Wood	Tree Range	Color of Heartwood	Color of Sapwood	Pattern Figure	Warpage	Strength	Uses
Alder, red (Alnus rubra)	Pacific coast	Light pinkish brown to white	Same	Obscure	Minor	Medium	Panel cores, table tops, sides, drawer fronts exposed parts of kitchen furniture. Stains readily in imitation of mahogany or walnut.
Ash, green, black, white (Fraxinus)	Eastern United States	Light grayish brown	White	Pronounced	Minor	High in-bending	Solid tables, dressers, wardrobes; wooden refrigerators.
Beech (Fagus grandifolia)	Eastern United States	White to slightly reddish	Same	Obscure	Pronounced	High	Chairs and exterior parts of painted furniture. Bends easily and is well adapted for curved parts such as chair backs. Also used for sides, guides, and backs of drawers, and for other substantial interior parts.
Birch, yellow and black (Betula)	Eastern United States	Light to dark reddish brown	White	Varying from a stripe to curly	Minor	High	Solid and veneered furniture. Same uses as hard maple.
Cherry, black (Prunus serotina)	Eastern United States	Light to dark reddish brown	White	Obscure	Minor	Medium	Solid furniture. Relative scarcity causes it to be quite expensive
Chestnut (Castanea dentata)	Eastern United States (now practically all blight-killed)	Grayish brown	White	Conspicuous	Minor	Medium	Cores of tables and dresser tops, drawer fronts, and other veneered panels. Used with oak in solid furniture.
Elm, American and rock (Ulmus)	Eastern United States	Light grayish brown often tinted with red	White	Conspicuous	Pronounced	High	Used to some extent for exposed parts of high-grade upholstered furniture. Easily bent to curved shapes such as chair backs.
Gum, red (Liquidambar styraciflua)	Eastern United States, Mexico and Guatemala	Reddish brown	Pinkish white	Obscure to figured	Minor	Medium	Gum furniture may be stained to resemble walnut or mahogany. Also used in combination with these woods.
Gum, tupelo (Nyssa aquatica)	Southeastern United States	Pale brownish gray	White	Obscure to striped	Pronounced	Medium	Cores of veneered panels, interior parts, framework of upholstered articles.

Name	Where grown	Heartwood color	Sapwood color	Figure	Shrinkage	Hardness	Uses
Mahogany (Swietenia, Khaya)	Mexico, Central America, West Indies, Africa	Pale to deep reddish brown	White to light brown	Ribbon or stripe	Minor	Medium	All solid and veneered high-grade furniture, boat construction, and cabinet work.
Maple, hard (Acer saccharum)	Eastern United States	Light reddish brown	White	Obscure to figured	Minor	High	Bedroom, kitchen, dining, and living room solid furniture. Some veneer (highly figured) is used. Mose furniture is given a natural finish.
Oak, red and white (Quercus)	Eastern United States	Grayish brown	White	Conspicuous	Minor	High	Solid and veneered furniture of all types. Quartered-oak furniture compares favorably with walnut and mahogany pieces and is often preferred in offices.
Pine, ponderosa (Pinus ponderosa)	Western United States	Light reddish	White	Obscure	Minor	Medium	Painted kitchen furniture.
Poplar, yellow (Liriodendron tulipifera)	Eastern and Southern United States	Light yellow to dark olive	White	Obscure	Slight	Medium	Cross banding of veneers, inexpensive painted furniture, interior portions of more expensive furniture, frames of upholstered articles.
Rosewood (Dalbergia nigra)	Eastern Brazil	Dark reddish brown with black streaks	White	Obscure streaked	Slight	High	Piano cases, musical instruments, handles, and so on.
Sycamore (Platanus occidentalis)	Eastern United States	Reddish brown	Pale reddish brown	Obscure to flake	Pronounced	High	Drawer sides, interior parts, frame work of upholstered articles.
Tanquile (Shorea)	Philippine Islands	Pale to dark reddish brown	Pale grayish to reddish brown	Ribbon or stripe	Slight	Medium	Similar to true mahogany.
Walnut, black (Juglans nigra)	Eastern United States	Light to dark chocolate brown	Pale brown	Varying from a stripe to a wave	Minor	High	All types of solid and veneered furniture.

CHAPTER 3

FURNITURE CONSTRUCTION

9. Frame Detail

Figures 4 to 28 show frame details for a few representative types of household furniture; construction details for office furniture are similar. Tables II to XXVIII, inclusive, list the materials used in the furniture illustrated.

TABLE II. Bill of materials for settee—three-seat

Key*	PART	No.	Length	Width	Thickness	Material	Remarks
			Dimensions (in.)				
1	Legs.....................	4	12½	1⅞	1⅞	Maple or birch	
2	Back posts................	4	35½	4½	1⅞		
3	Front rail—seat............	1	69	3½	1¼		
4	Side rail—seat.............	2	24¼	3½	1¼		
5	Back rail—seat.............	3	22	3½	1¼		
6	Stretchers.................	4	25	1⅝	1⅛		
7	Stretchers.................	3	22	1⅝	1⅛		
8	Arm stumps................	2	14½	3	1¾		
9	Arms.....................	2	26	4	1½		
10	Back rails.................	9	22	2	⅞		
11	Front rail—seat frame.......	1	69¼	3⅛	1¼	Hardwood	
12	Side rail—seat frame........	2	23	3⅛	1¼		
13	Back rail—seat frame.......	1	64½	3⅛	1¼		
14	Front cleat................	1	66	1½	1¼		
15	Side cleat.................	2	21½	1½	1¼		
16	Back cleat.................	1	63	1½	1¼		
17	Center brace..............	1	21¾	3¼	2½		

*See figure 4.

TABLE III. Bill of materials for settee—two-seat

Key*	PART	No.	Length	Width	Thickness	Material	Remarks
			Dimensions (in)				
1	Legs.....................	3	12½	1⅞	1⅞	Maple or birch	
2	Back posts................	3	35½	4½	1⅞		
3	Front rail—seat............	1	45¼	3½	1¼		
4	Side rail—seat.............	2	24¼	3½	1¼		
5	Back rails—seat............	2	20½	3½	1¼		
6	Stretchers.................	3	25	1⅝	1⅛		
7	Stretchers.................	2	22	1⅝	1⅛		
8	Arm stumps................	2	14½	3	1¾		
9	Arms.....................	2	26	4	1½		
10	Back rails.................	6	22	2	⅞		
11	Front rail—seat frame.......	1	42½	3⅛	1¼	Hardwood	
12	Side rail—seat frame........	2	23	3⅛	1¼		
13	Back rail—seat frame.......	1	41	3⅛	1¼		
14	Front cleat................	1	43	1½	1¼		
15	Side cleats................	2	21½	1½	1¼		
16	Back cleats................	1	40	1½	1¼		
17	Center brace..............	1	21¾	3¼	2½		

*See figure 4.

TABLE IV. Bill of materials for easy chair

Key*	PART	No.	Dimensions (in.) Length	Width	Thickness	Material	Remarks
1	Legs	2	12½	1⅞	1⅞	Maple or birch	Turned.
2	Back posts	2	35½	4½	1⅞	Maple or birch	
3	Front rail—seat	1	22⅜	3½	1¼	Maple or birch	
4	Side rail—seat	2	24¼	3½	1¼	Maple or birch	
5	Back rail—seat	1	19¾	3½	1¼	Maple or birch	
6	Stretchers	2	25	1⅝	1⅛	Maple or birch	
7	Stretchers	1	22	1⅝	1⅛	Maple or birch	
8	Arm stumps	2	14½	3	1¾	Maple or birch	
9	Arms	2	26	4	1½	Maple or birch	
10	Back rails	3	21	2	⅞	Maple or birch	
11	Front rail—seat frame	1	23¼	3⅛	1¼	Hardwood	
12	Side rail—seat frame	2	21½	3⅛	1¼	Hardwood	
13	Back rail—seat frame	1	18	3⅛	1¼	Hardwood	
14	Front cleat	1	20¼	1½	1¼	Hardwood	
15	Side cleat	2	21½	1½	1¼	Hardwood	
16	Back cleat	1	18	1½	1¼	Hardwood	

*See figure 4.

TABLE V. Bill of materials for occasional armchair

Key*	PART	No.	Dimensions (in.) Length	Width	Thickness	Material	Remarks
1	Back posts	2	34¼	5⅜	1 1/16	Maple or birch	
2	Front legs	2	13⅞	1 9/16	1 9/16	Maple or birch	Turned.
3	Arm stumps	2	10½	1 5/16	1 5/16	Maple or birch	Turned.
4	Arms	2	19½	2⅜	1 5/16	Maple or birch	Shaped.
5	Back rail	1	17⅜	2¼	13/16	Hardwood	
6	Front rail	1	20	2¼	13/16	Hardwood	
7	Side rails	2	20½	2¼	13/16	Hardwood	
8	Back slats	2	18	2	9/16	Maple or birch	
9	Side stretchers	2	22½	1 5/16	9/16	Maple or birch	
10	Center stretcher	1	20⅛	1 5/16	9/16	Maple or birch	
11	Back slat	1	18	3¼	9/16	Maple or birch	
	Spring seat						
	Pad back						

*See figure 5.

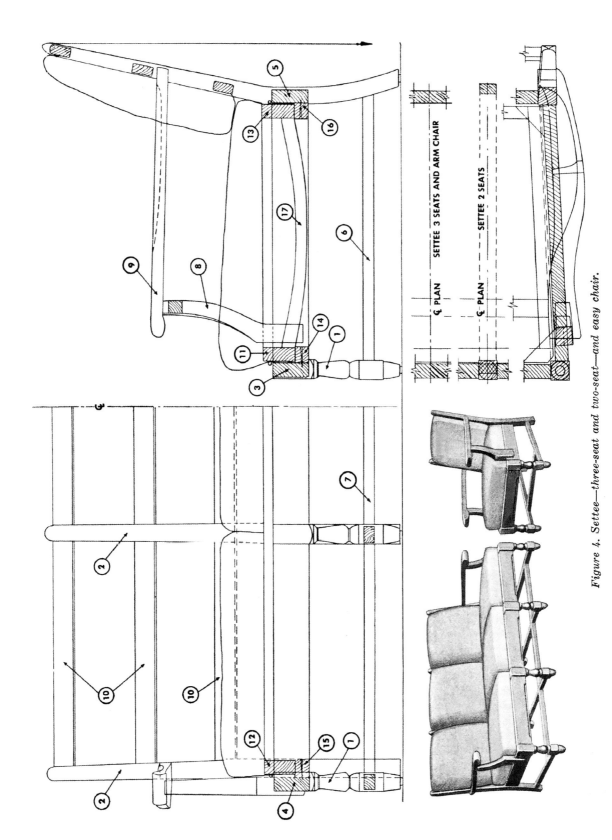

Figure 4. Settee—three-seat and two-seat—and easy chair.

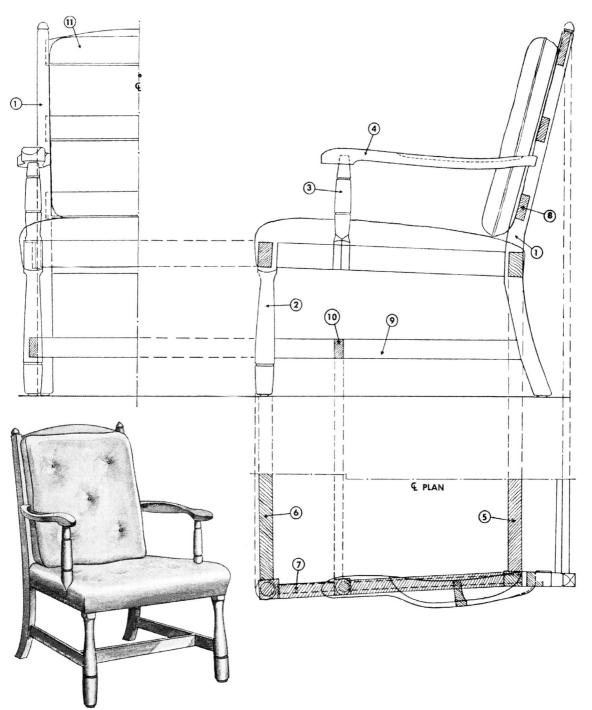

Figure 5. Occasional armchair.

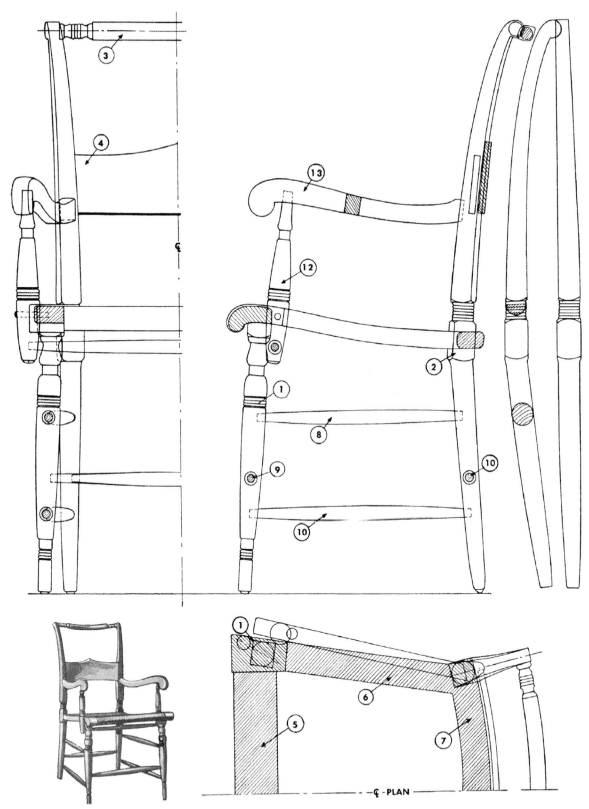

Figure 6. Occasional or dining armchair—rush seat.

TABLE VI. *Bill of materials for occasional or dining armchair—rush seat*

Key*	PART	No.	Length	Width	Thickness	Material	Remarks
			Dimensions (in.)				
1	Front legs.................	2	17	$1\frac{9}{16}$	$1\frac{9}{16}$		Turned.
2	Back posts.................	2	36	$3\frac{3}{4}$	$1\frac{9}{16}$	Maple or birch	Turned and shaped.
3	Top rail...................	1	17	2	$1\frac{1}{16}$		Turned and shaped.
4	Splat.....................	1	$15\frac{1}{2}$	$4\frac{5}{8}$	$1\frac{1}{4}$		Shaped.
5	Front rail.................	1	$15\frac{1}{2}$	$2\frac{7}{8}$	$1\frac{1}{2}$		Shaped.
6	Side rails.................	2	$11\frac{3}{4}$	$1\frac{7}{8}$	$1\frac{1}{2}$	Hardwood	Shaped.
7	Back rail..................	1	$13\frac{3}{4}$	$2\frac{1}{2}$	1		
8	Stretchers—side...........	2	$13\frac{3}{4}$	$\frac{7}{8}$	$\frac{7}{8}$		Turned.
9	Stretcher..................	1	$17\frac{1}{2}$	$\frac{7}{8}$	$\frac{7}{8}$		Turned.
10	Stretchers—side, back......	3	15	$\frac{7}{8}$	$\frac{7}{8}$	Maple or birch	Turned.
11	Blocks....................	2	4	$2\frac{1}{2}$	$1\frac{3}{4}$		Shaped.
12	Arm stumps...............	2	$11\frac{1}{2}$	$1\frac{1}{2}$	$1\frac{1}{2}$		Turned.
13	Arms.....................	2	15	$2\frac{5}{8}$	$1\frac{1}{16}$		Shaped.

*See figure 6.

TABLE VII. *Bill of materials for occasional or dining chair—rush seat*

Key*	PART	No.	Length	Width	Thickness	Material	Remarks
			Dimensions (in.)				
1	Back posts.................	2	$36\frac{3}{4}$	$4\frac{1}{2}$	$1\frac{5}{8}$		
2	Front legs.................	2	$17\frac{3}{4}$	$1\frac{5}{8}$	$1\frac{5}{8}$		Turned.
3	Front rail.................	1	$16\frac{3}{4}$	$2\frac{1}{4}$	$1\frac{3}{8}$		
4	Back rail..................	1	$12\frac{3}{4}$	$2\frac{1}{4}$	$1\frac{3}{8}$		
5	Side rail..................	2	$13\frac{7}{8}$	$2\frac{1}{4}$	$1\frac{3}{8}$		
6	Side stretchers.............	2	$15\frac{3}{8}$	$1\frac{1}{2}$	$\frac{3}{4}$	Maple or birch	
7	Center stretcher...........	1	17	$1\frac{1}{2}$	$\frac{3}{4}$		
8	Back stretcher.............	1	$11\frac{3}{4}$	$1\frac{1}{2}$	$\frac{3}{4}$		
9	Back rail—top.............	1	$16\frac{1}{8}$	$3\frac{1}{2}$	$1\frac{1}{2}$		
10	Back rail..................	1	16	$2\frac{5}{8}$	$\frac{5}{8}$		
11	Back rail..................	1	$15\frac{1}{4}$	$2\frac{5}{8}$	$\frac{5}{8}$		
12	Back rail..................	1	$14\frac{1}{2}$	$2\frac{5}{8}$	$\frac{5}{8}$		
13	Side rails—rush seat........	2	$11\frac{3}{8}$	$1\frac{3}{4}$	$\frac{3}{4}$		
14	Front rail—rush seat........	1	$13\frac{5}{8}$	$1\frac{3}{4}$	$\frac{3}{4}$	Hardwood	
15	Back rail—rush seat........	1	$10\frac{3}{8}$	$1\frac{3}{4}$	$\frac{3}{4}$		
16	Blocks—rush seat..........	4	$2\frac{1}{4}$	$2\frac{1}{4}$	$1\frac{1}{8}$	Maple or birch	

*See figure 7.

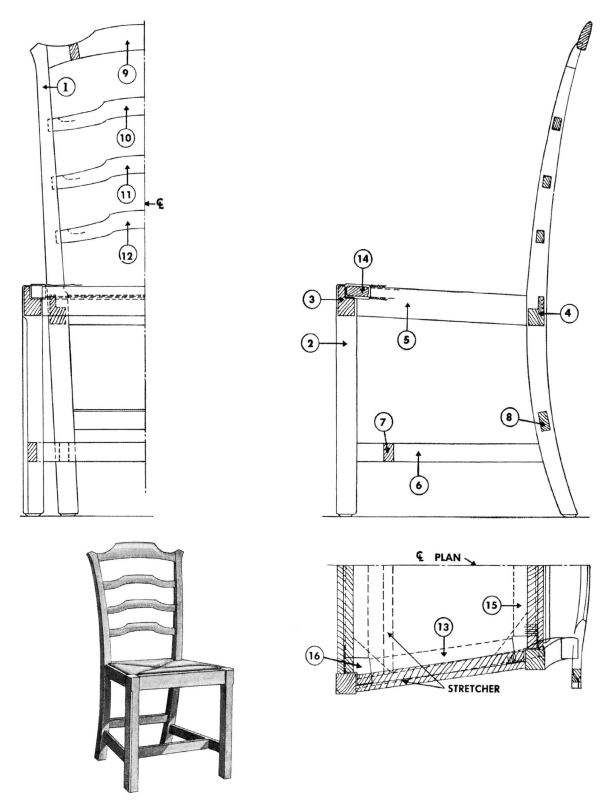

Figure 7. Occasional or dining chair—rush seat.

TABLE VIII. Bill of materials for dining chair—slip seat

Key*	PART	No.	Dimensions (in.)			Material	Remarks
			Length	Width	Thickness		
1	Front legs.................	2	17	1⅝	1⅝		Shaped.
2	Back posts................	2	34½	4⅛	1⅝		Shaped.
3	Top rail—back.............	1	19	2½	1½		Shaped.
4	Back splats................	2	13½	2	½		
5	Stretchers................	2	16	1⅜	1⅜	}Maple or birch	Turned.
6	Stretcher.................	1	15¼	1⅜	1⅜		Turned.
7	Front rail—seat...........	1	16	1⅞	1¼		
8	Side rails—seat...........	2	13¾	1⅞	1¼		
9	Back rail—seat...........	1	11½	1⅞	1¼		
10	Front rail—slip seat........	1	18	2⅛	⅞		
11	Side rail—slip seat..........	2	13½	2⅛	⅞	}Hardwood	
12	Back rail—slip seat..........	1	10	2⅛	⅞		
	Upholstered slip seat						

*See figure 8.

TABLE IX. Bill of materials for windsor armchair—wood

Key*	PART	No.	Dimensions (in.)			Material	Remarks
			Length	Width	Thickness		
1	Back rail.................	1	37	¾	¾		Steam-bent.
2	Arm rail.................	1	17	4	13/16		
3	Arms....................	2	14½	2½	13/16		
4	Spindles.................	6	20	¾	¾		Turned.
5	Spindles.................	2	15	1⅝	¾		Turned.
6	Spindles.................	2	10	¾	¾	}Maple or birch	Turned.
7	Arm stumps..............	2	10¾	1⅝	1⅝		Turned.
8	Seat....................	1	18	19	1 5/16		
9	Legs....................	4	18	1⅞	1⅞		Turned.
10	Side stretchers............	2	19	1 7/16	1 7/16		Turned.
11	Front stretcher.............	1	16½	1½	1½		Turned.
12	Back stretcher.............	1	13½	1¼	1¼		Turned.

*See figure 9.

TABLE X. Bill of materials for windsor side chair—wood

Key*	PART	No.	Dimensions (in.)			Material	Remarks
			Length	Width	Thickness		
1	Back rail.................	1	54	⅞	⅞		Steam-bent.
2	Spindles..................	10	20	⅝	⅝		Turned.
3	Seat....................	1	17¼	18⅝	1¼	}Maple or birch	Shaped.
4	Legs....................	4	17¼	1⅝	1⅝		Turned.
5	Side stretchers............	2	15½	1⅜	1⅜		Turned.
6	Center stretcher............	1	15¾	1	1		Turned.

*See figure 10.

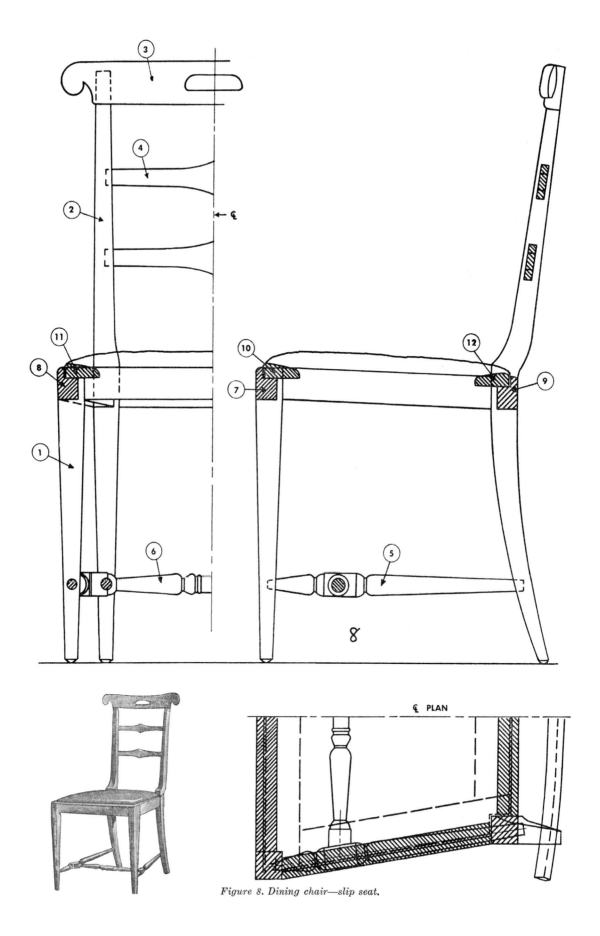

Figure 8. Dining chair—slip seat.

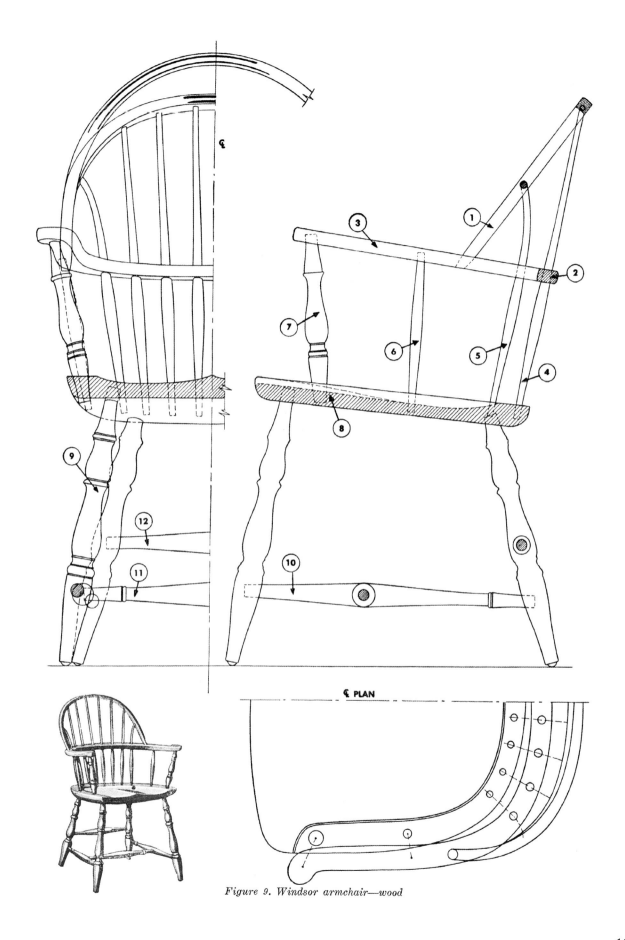

Figure 9. Windsor armchair—wood

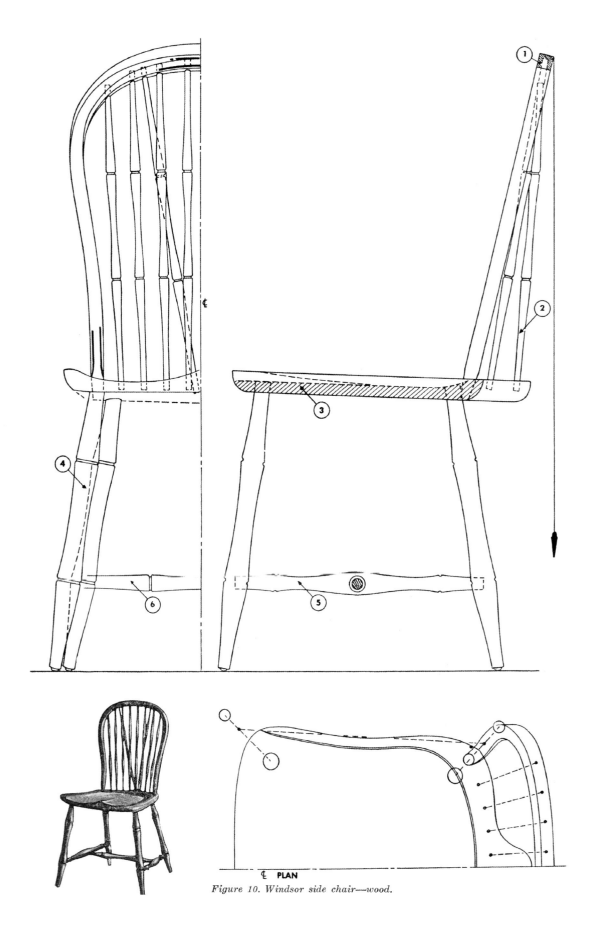

Figure 10. Windsor side chair—wood.

TABLE XI. Bill of materials for slant-top desk and secretary desk

Key*	PART	No.	Dimensions (in.)			Material	Remarks
			Length	Width	Thickness		
1	Top panel (desk)...........	1	31	9¼	¹³⁄₁₆	⎫	
2	Side panels...............	2	39¼	19⅛	¹³⁄₁₆	⎬ Maple or birch	
3	Front rail................	1	29¼	2½	1⅛	⎭	
4	Back rail.................	1	29¼	1¼	1⅛	Hardwood	
5	Desk top.................	1	30	18½	¹³⁄₁₆	⎫	
6	Droplid panel.............	1	26½	13⅛	¹³⁄₁₆	⎬ Maple or birch	
7	Droplid end rail...........	1	13⅛	2⅛	¹³⁄₁₆	⎭	
8	Front rails...............	3	30	2¼	¹³⁄₁₆	⎫	
9	Back rails................	3	30	2¼	¹³⁄₁₆	⎪	
10	End rails.................	6	14⅝	2¼	¹³⁄₁₆	⎬ Hardwood (3-ply)	
11	Dust bottoms.............	3	26	14⅝	³⁄₁₆	⎭	
12	Drawer front.............	1	28	5	¹³⁄₁₆	⎫	
13	Drawer front.............	1	30	6	¹³⁄₁₆	⎬ Maple or birch	
14	Drawer front.............	1	30	7	¹³⁄₁₆	⎭	
15	Drawer sides..............	2	17⅞	5	½	⎫	
16	Drawer sides..............	2	17⅞	6	½	⎪	
17	Drawer sides..............	2	17⅞	7	½	⎪	
18	Drawer back..............	1	27	5	½	⎪	
19	Drawer back..............	1	29	6	½	⎬ Hardwood (3-ply)	
20	Drawer back..............	1	29	7	½	⎪	
21	Drawer bottoms...........	3	17½	28⅜	³⁄₁₆	⎪	
22	Drawer slides.............	3	17½	2¼	⁷⁄₁₆	⎪	
23	Drawer guides............	3	16¾	1	⁹⁄₁₆	⎭	
24	Basefront................	1	31½	7½	¹³⁄₁₆	⎫	
25	Pigeonhole dividers........	6	7⅛	8	¼	⎪	
26	Pigeonhole top............	1	29¼	8¾	¼	⎪	
27	Sidescenter drawer.........	2	8	5¾	⅜	⎪	
28	Pilasters.................	2	5¾	¾	½	⎬ Maple or birch	
29	Guides..................	2	7⅛	⅞	⁵⁄₁₆	⎪	
30	Backs...................	2	11¾	5¾	¼	⎪	
31	Shelves..................	2	11¼	6⅛	¼	⎪	
32	Bottoms.................	2 and 1	8⅝–13	8½	¼	⎭	
33	Drawer knobs.............	6	1⅝	1⅝	1		
34	Droplid hinges............	2				Keeler No. 4589	Or equal.
35	Back panel...............	1	32	30	³⁄₁₆	Hardwood (3-ply)	
36	Lock....................	1					
36–A	Slide supports.............	2	17⅞	5	¹³⁄₁₆	⎱ Maple or birch	Turned.
36–B	Pulls...................	2	⅝	⅝	½	⎰	

*See figure 11.

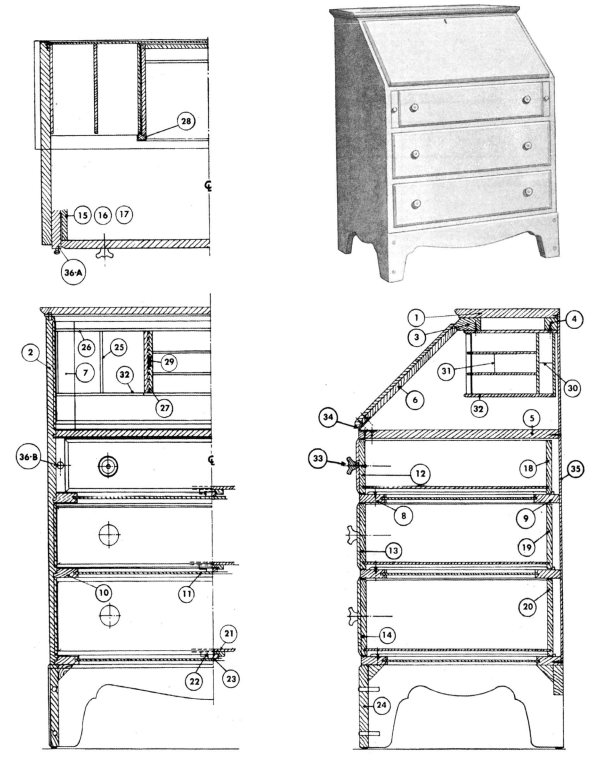

Figure 11. Slant-top desk and secretary desk.

TABLE XII. Bill of materials for secretary desk

Key*	PART	No.	Dimensions (in.)			Material	Remarks
			Length	Width	Thickness		
37	Top panel (upper part).......	1	$30\frac{3}{8}$	9	$\frac{13}{16}$		
38	Rail......................	1	29	$1\frac{1}{8}$	$1\frac{1}{8}$		
39	Sides.....................	2	$30\frac{1}{2}$	$8\frac{7}{8}$	$\frac{13}{16}$		
40	Bottom...................	1	$31\frac{1}{2}$	$9\frac{3}{8}$	$\frac{13}{16}$		
41	Stop strip................	1	29	$\frac{1}{2}$	$\frac{1}{4}$		
42	Door stiles................	2	$29\frac{5}{8}$	$1\frac{3}{8}$	$\frac{13}{16}$	Maple or birch	
43	Door stiles................	2	$29\frac{5}{8}$	$1\frac{5}{8}$	$\frac{13}{16}$		
44	Door rails.................	4	$12\frac{1}{4}$	$1\frac{5}{8}$	$\frac{13}{16}$		
45	Muntins...................	2	27	$\frac{1}{2}$	$\frac{3}{8}$		
46	Muntins...................	4	$11\frac{1}{2}$	$\frac{1}{2}$	$\frac{3}{8}$		
47	Shelves—fixed.............	2	29	$7\frac{3}{4}$	$\frac{1}{2}$		
48	Back panel................	1	$31\frac{3}{8}$	$29\frac{3}{4}$	$\frac{3}{16}$	Hardwood (3-ply)	
49	Hinges....................	4				Keeler No. 6832	Or equal.
50	Lock......................	1					
51	Glass panes................	2	$27\frac{1}{2}$	$12\frac{1}{4}$			

*See figure 12.

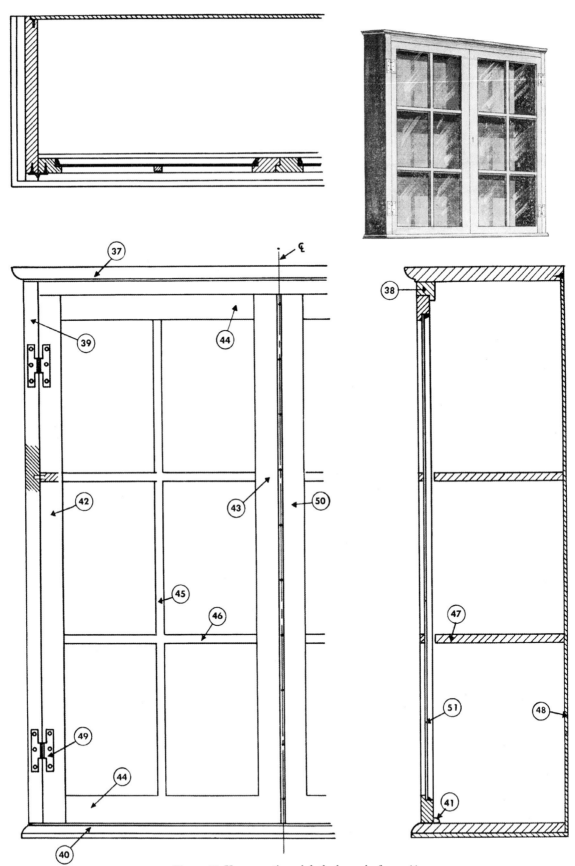

Figure 12. Upper section of desk shown in figure 11.

TABLE XIII. *Bill of materials for chest desk*

Key*	PART	No.	Dimensions (in.)			Material	Remarks
			Length	Width	Thickness		
1	Top panel	1	32	20	13/16		
2	End panels	2	30 1/16	18 3/4	13/16		
3	Front rails	5	29 5/8	2 1/4	13/16	} Maple or birch	
4	Back rails	5	29 5/8	2 1/4	13/16		
5	End rails	10	14 3/4	2 1/4	13/16		
6	Dust bottoms	4	25 7/8	14 3/4	3/16	3-ply hardwood	
7	Center glides	3	17 5/8	2 1/4	7/16	} Hardwood	
8	Center guides	4	17 1/2	1	9/16		
9	Drawer front	1	29 3/8	5	13/16		
10	Drawer front	1	29 3/8	6	13/16	} Maple or birch	
11	Drawer front	1	29 3/8	7	13/16		
12	Drawer front	1	29 3/8	8	13/16		
13	Drawer sides	2	18	4 7/8	13/16		
14	Drawer sides	2	18	5 7/8	1/2		
15	Drawer sides	2	18	6 7/8	1/2		
16	Drawer sides	2	18	7 7/8	1/2		
17	Drawer back	1	28 5/8	4 7/8	1/2	} Hardwood	
18	Drawer back	1	28 5/8	5 7/8	1/2		
19	Drawer back	1	28 5/8	6 7/8	1/2		
20	Drawer back	1	28 5/8	7 7/8	1/2		
21	Drawer bottom	1	27 1/2	17 3/8	7/16		
22	Drawer bottoms	3	28 1/8	17 7/8	3/16	3-ply hardwood	
23	Front base	1	31 5/8	6	1 1/16		
24	End bases	2	19 1/4	6	1 1/16		
25	Back base	1	29 1/2	4 1/2	13/16		
26	Pigeonhole shelf	1	27	7 3/4	1/4	} Maple or birch	
27	Pigeonhole top	1	27	7 3/4	1/4		
28	Pigeonhole verticals	4	4 1/4	7 1/2	1/4		
29	Pigeonhole drawer front	1	12	1 7/8	1/2		
30	Pigeonhole drawer sides	2	7 1/4	1 13/16	3/8	} Hardwood	
31	Pigeonhole drawer back	1	12	1 13/16	3/8		
32	Pigeonhole drawer bottom	1	11 1/2	6 7/8	1/8	3-ply hardwood	
33	Pigeonhole drawer knobs	2	3/4	3/4	13/16	} Maple or birch	Turned.
34	Drawer knobs	8	1 3/4	1 3/4	15/16		Turned.
35	Fall support	1				Keeler No. 3793	Or equal.
36	Case back	1	30 1/16	29 3/4	3/16	3-ply hardwood	

*See figure 13.

TABLE XIV. Bill of materials for kneehole desk

Key*	PART	No.	Length	Width	Thickness	Material	Remarks
			Dimensions (in.)				
1	Top panel.................	1	45	24	$\frac{13}{16}$		
2	End panels................	4	$22\frac{7}{16}$	$22\frac{3}{4}$	$\frac{13}{16}$		
3	Back panels...............	2	$22\frac{7}{16}$	11	$\frac{13}{16}$		
4	Bases—front and back.......	4	$12\frac{1}{8}$	$4\frac{3}{4}$	$1\frac{1}{16}$	Maple or birch	
5	Bases—ends...............	4	$23\frac{7}{8}$	$4\frac{3}{4}$	$1\frac{1}{16}$		
6	Front rails................	8	$10\frac{1}{8}$	$2\frac{1}{4}$	$\frac{13}{16}$		
7	Back rails.................	8	$10\frac{1}{8}$	$2\frac{1}{4}$	$\frac{13}{16}$	Hardwood	
8	Side rails.................	16	$18\frac{5}{8}$	$2\frac{1}{4}$	$\frac{13}{16}$		
9	Dust bottoms..............	6	$18\frac{1}{2}$	$6\frac{3}{8}$	$\frac{3}{16}$	3-ply hardwood	
10	Drawer fronts.............	2	$9\frac{3}{8}$	5	$\frac{13}{16}$		
11	Drawer fronts.............	2	$9\frac{3}{8}$	7	$\frac{13}{16}$	Maple or birch	
12	Drawer fronts.............	2	$9\frac{3}{8}$	8	$\frac{13}{16}$		
13	Drawer sides.............	4	$21\frac{1}{4}$	$4\frac{7}{8}$	$\frac{1}{2}$		
14	Drawer sides.............	4	$21\frac{1}{4}$	$6\frac{7}{8}$	$\frac{1}{2}$		
15	Drawer sides.............	4	$21\frac{1}{4}$	$7\frac{7}{8}$	$\frac{1}{2}$	Hardwood	
16	Drawer backs.............	2	$9\frac{1}{8}$	$4\frac{7}{8}$	$\frac{1}{2}$		
17	Drawer backs.............	2	$9\frac{1}{8}$	$6\frac{7}{8}$	$\frac{1}{2}$		
18	Drawer backs.............	2	$9\frac{1}{8}$	$7\frac{7}{8}$	$\frac{1}{2}$		
19	Drawer bottoms............	6	$20\frac{7}{8}$	$8\frac{1}{2}$	$\frac{3}{16}$	3-ply hardwood	
20	Drawer slides.............	6	$20\frac{3}{4}$	$2\frac{1}{4}$	$\frac{7}{16}$	Hardwood	
21	Drawer guides............	6	$20\frac{1}{2}$	1	$\frac{9}{16}$		
22	Neck molds...............	4	$11\frac{5}{8}$	$\frac{7}{16}$	$\frac{1}{4}$		
23	Neck molds...............	4	$23\frac{3}{8}$	$\frac{7}{16}$	$\frac{1}{4}$		
24	Base molds...............	4	12	$\frac{13}{16}$	$\frac{7}{16}$	Maple or birch	
25	Base molds...............	4	$23\frac{3}{4}$	$\frac{13}{16}$	$\frac{7}{16}$		
26	Aprons...................	2	22	$4\frac{1}{2}$	$\frac{13}{16}$		
27	Drawer pulls..............	6	5	$\frac{13}{16}$	$\frac{13}{16}$		

*See figure 14.

TABLE XV. Bill of materials for bookshelves—five shelves

Key*	PART	No.	Length	Width	Thickness	Material	Remarks
			Dimensions (in.)				
1	Shelf.....................	1	$34\frac{1}{4}$	$7\frac{3}{8}$	$\frac{13}{16}$		
2	Shelf.....................	1	$34\frac{1}{4}$	$8\frac{1}{2}$	$\frac{13}{16}$		
3	Shelf.....................	1	$34\frac{1}{4}$	9	$\frac{13}{16}$		
4	Shelf.....................	1	$34\frac{1}{4}$	$9\frac{1}{2}$	$\frac{13}{16}$		
5	Bottom shelf..............	1	37	$11\frac{1}{4}$	$\frac{13}{16}$		
6	Sides.....................	2	$47\frac{1}{8}$	11	$\frac{13}{16}$	Maple or birch	
7	Front base................	1	37	$3\frac{7}{8}$	$\frac{13}{16}$		
8	End bases................	2	$11\frac{1}{2}$	$3\frac{7}{8}$	$\frac{13}{16}$		
9	Cross brace...............	1	36	$2\frac{1}{2}$	$\frac{13}{16}$		
10	Back panel................	1	47	$35\frac{1}{4}$	$\frac{3}{16}$		Plywood.
11	Cross brace...............	1	$34\frac{1}{2}$	$3\frac{7}{8}$	$\frac{13}{16}$		

*See figure 15.

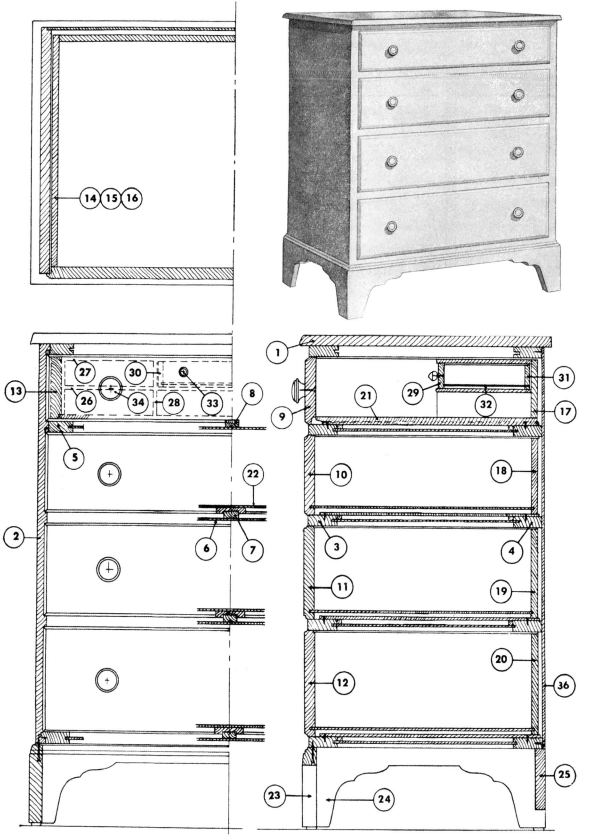

Figure 13. Chest desk.

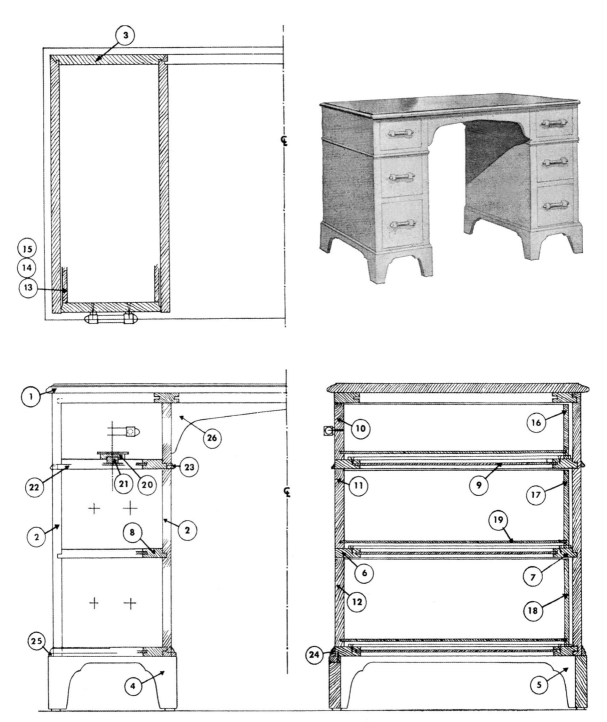

Figure 14. Kneehole desk.

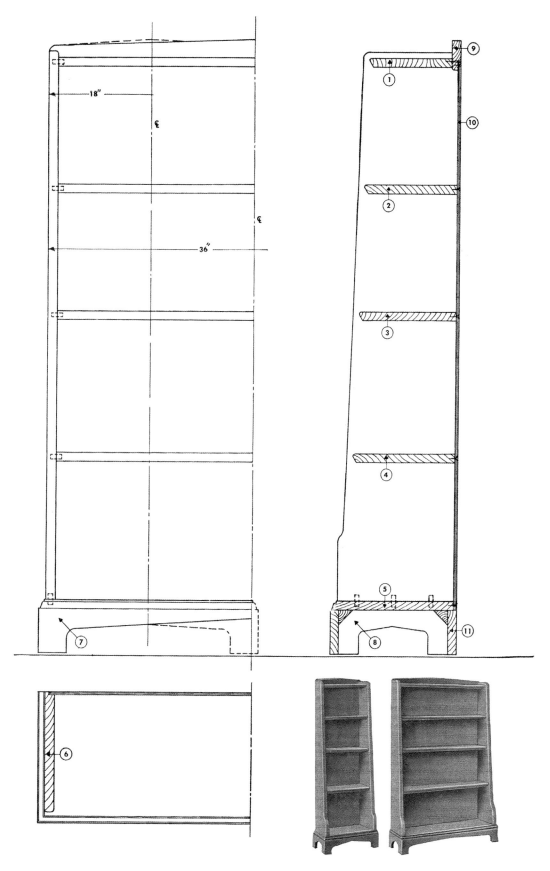

Figure 15. Bookshelves.

TABLE XVI. *Bill of materials for sofa or wall table*

Key*	PART	No.	Dimensions (in.)			Material	Remarks
			Length	Width	Thickness		
1	Top panel.................	1	60	28	$\frac{7}{8}$		Turned.
2	Legs......................	4	29		$2\frac{3}{8}$ x $2\frac{3}{8}$		
3	End aprons...............	2	$19\frac{1}{4}$	4	$1\frac{1}{8}$		
4	Side aprons..............	2	$44\frac{1}{4}$	4	$1\frac{1}{8}$	Maple or birch	
5	Brackets.................	8	$5\frac{5}{8}$	$2\frac{5}{8}$	$1\frac{1}{8}$		
6	End stretchers...........	2	$19\frac{1}{4}$	$2\frac{1}{4}$	$1\frac{1}{4}$		
7	Center stretcher..........	1	$45\frac{3}{8}$	$2\frac{1}{4}$	$1\frac{1}{4}$		

*See figure 16.

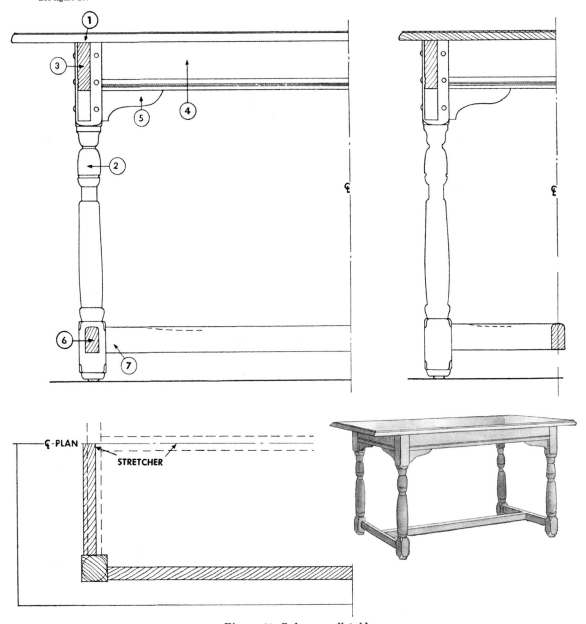

Figure 16. Sofa or wall table.

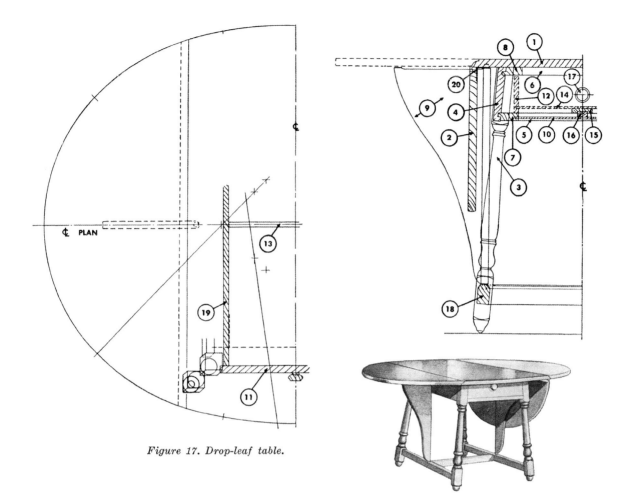

Figure 17. Drop-leaf table.

TABLE XVII. Bill of materials for drop-leaf table

Key*	PART	No.	Length	Width	Thickness	Material	Remarks
			Dimensions (in.)				
1	Top—stationary.............	1	42	24¾	¹³⁄₁₆		
2	Drop leaves.................	2	39½	15⅝	¹³⁄₁₆	Maple or birch	
3	Legs......................	4	28¼	1¹³⁄₁₆	1¹³⁄₁₆		Turned.
4	Aprons....................	2	28⅛	5¾	¹³⁄₁₆		
5	Rails—bottom ends.........	2	18⅝	2¼	¹³⁄₁₆	Hardwood	
6	Rails—top ends............	2	18	2¼	¹³⁄₁₆		
7	Rails—bottom sides........	2	27¾	2¼	¹³⁄₁₆		
8	Rails—top sides............	2	26¾	2¼	¹³⁄₁₆	Maple or birch	
9	Brackets...................	2	22¾	10⅝	¹³⁄₁₆		
10	Dust panel................	1	27¾	15	³⁄₁₆	3-ply hardwood	
11	Drawer fronts.............	2	16¼	4	¹³⁄₁₆	Maple or birch	
12	Drawer sides..............	2	30½	3⅞	½	Hardwood	
13	Drawer center............	1	14	3³⁄₁₆	½		
14	Drawer bottom............	1	30¼	14¾	³⁄₁₆	3-ply hardwood	
15	Drawer slide..............	1	29¾	2¼	⁷⁄₁₆	Hardwood	
16	Drawer guide.............	1	29	1	⁹⁄₁₆		
17	Knobs....................	2	1⁹⁄₁₆	1⁹⁄₁₆	¹³⁄₁₆		Turned.
18	Stretchers.................	2	31¾	2⅛	1⁵⁄₁₆	Maple or birch	
19	Stretcher—center..........	1	20¼	2⅛	1⁵⁄₁₆		
20	Hinges...................	4					

*See figure 17.

TABLE XVIII. Bill of materials for dining draw table with stretcher bars

Key*	PART	No.	Dimensions (in.)			Material	Remarks
			Length	Width	Thickness		
1	Top panel.................	1	60½	38	⅞		
2	Top end cleats............	2	38	3¼	⅞		
3	Top end inlays............	6	2	1¼	¼	} Maple or birch	
4	Draw leaves...............	2	38	21	⅞		
5	Filler pieces...............	2	24	4¼	⅞		
6	Slides....................	4	54½	3	1⅜		
7	Slide tie.................	1	19¼	2	1⅛	} Hardwood	
8	Bridge...................	1	29⅝	2½	⅞		
9	Bridge...................	1	31¼	2½	1¼		
10	Aprons...................	2	48	3¾	1¼		
11	Aprons...................	2	30	3¾	1¼		
12	Brackets.................	8	5	2½	1¼	} Maple or birch	
13	Legs.....................	4	28½	2¼	2¼		
14	Stretchers................	4	23	2⅛	1¼		
15	Stretcher.................	1	17½	2⅛	1¼		
16	Metal plates..............	4	2½	⅞	⅛	Iron	Slide clips.

*See figure 18.

TABLE XIX. Bill of materials for dining draw table

Key*	PART	No.	Dimensions (in.)			Material	Remarks
			Length	Width	Thickness		
1	Top panel.................	1	54	56	13/16		
2	Draw leaves...............	2	36	16	13/16		
3	Slides....................	4	43	3⅛	1¼		
4	Aprons...................	2	42⅝	3⅜	1⅛		
5	Aprons...................	2	28	3⅜	1⅛	} Maple or birch	
6	Brackets.................	8	2¼	2⅜	1⅛		
7	Bridge...................	1	30	1½	13/16		
8	Bridge...................	1	30	2⅝	1⅛		
9	Filler pieces...............	2	12	3	13/16		
10	Legs.....................	4	28⅛	2¼	2¼		Turned.
11	Metal plates..............	4	1¾	⅞	⅛	Iron	Slide clips.
12	Leg holders...............	4					

*See figure 19.

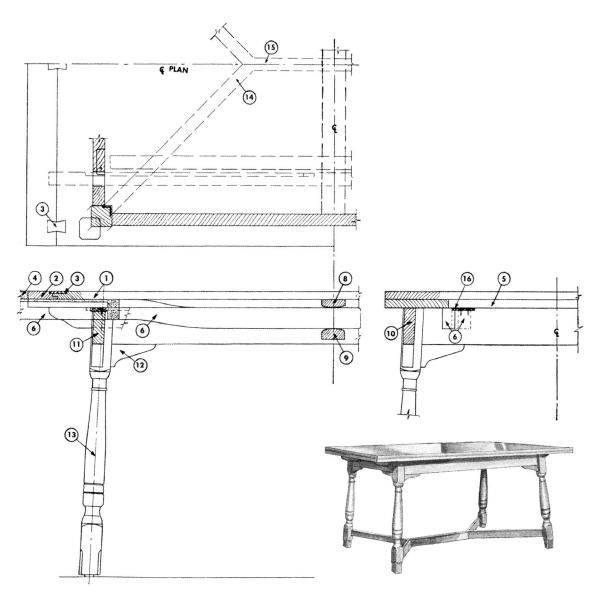

Figure 18. Dining draw table with stretcher bars.

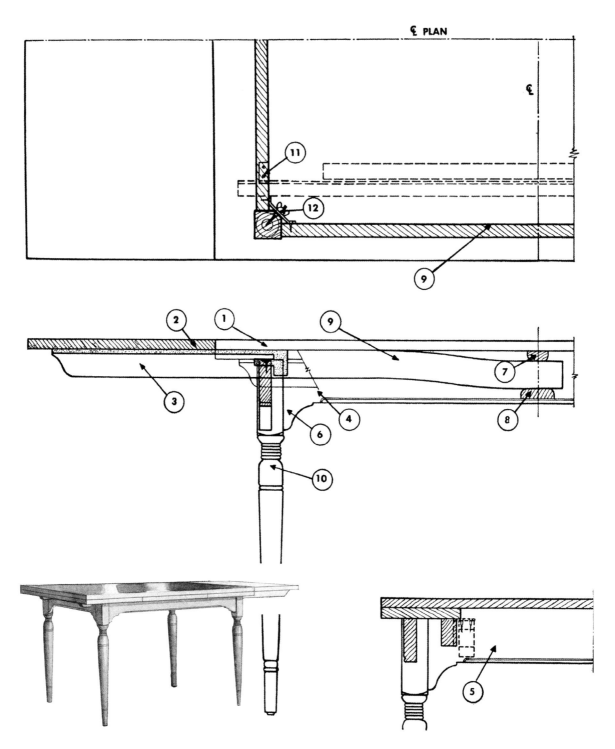

Figure 19. Dining draw table.

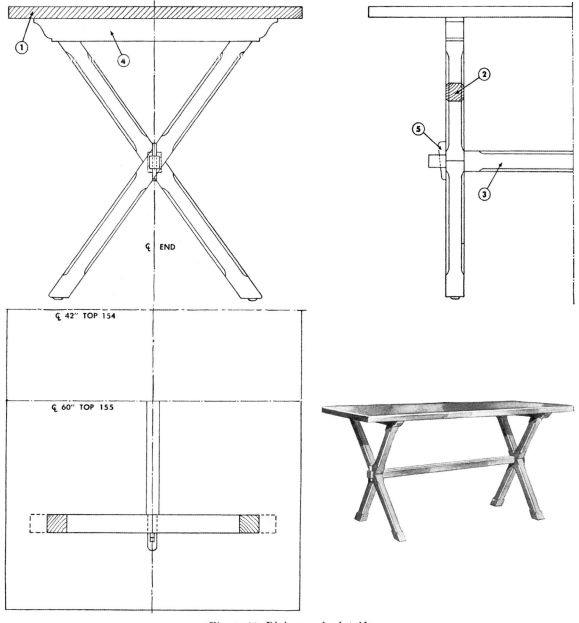

Figure 20. Dining sawbuck table.

TABLE XX. *Bill of materials for dining sawbuck table*

Key*	PART	No.	Dimensions (in.)			Material	Remarks
			Length	Width	Thickness		
1	Top panel..................	1	60	30	$1\frac{1}{8}$	⎫	
2	Legs......................	4		$1\frac{13}{16}$	$1\frac{13}{16}$	⎪	
3	Stretcher..................	1	$47\frac{1}{2}$	2	$1\frac{5}{16}$	⎬ Maple or birch	
4	Cross braces...............	2	25	$2\frac{1}{4}$	$1\frac{13}{16}$	⎪	
5	Keys......................	2	$3\frac{3}{4}$	$\frac{7}{8}$	$\frac{3}{8}$	⎭	

*See figure 20.

TABLE XXI. Bill of materials for sideboard

Key*	PART	No.	Dimensions (in.)			Material	Remarks
			Length	Width	Thickness		
1	Top panel.................	1	60	$20\frac{1}{2}$	$\frac{13}{16}$		
2	Legs......................	4	37	$1\frac{13}{16}$	$1\frac{13}{16}$		Turned.
3	End panels.................	2	$23\frac{1}{16}$	$15\frac{3}{8}$	$\frac{13}{16}$	Maple or birch	
4	Compartment sides..........	2	$15\frac{13}{16}$	$18\frac{1}{16}$	$\frac{13}{16}$		
5	Drawer stiles..............	2	4	$2\frac{1}{4}$	$\frac{13}{16}$		
6	Front rails.................	5	55	$2\frac{1}{4}$	$\frac{13}{16}$		
7	Back rails..................	5	55	$2\frac{1}{4}$	$\frac{13}{16}$		
8	End rails...................	8	$14\frac{3}{8}$	$2\frac{1}{4}$	$\frac{13}{16}$	Hardwood	
9	Center rails................	8	$14\frac{3}{8}$	$2\frac{1}{4}$	$\frac{13}{16}$		
10	End stretchers.............	2	$14\frac{3}{4}$	$1\frac{3}{4}$	$1\frac{1}{16}$	Maple or birch	
11	Long stretcher.............	1	$55\frac{3}{8}$	$1\frac{3}{4}$	$1\frac{1}{16}$		
12	Dust bottoms..............	2	11	$14\frac{1}{4}$	$\frac{3}{16}$	3-ply hardwood	
13	Dust bottoms..............	3	$26\frac{1}{4}$	$14\frac{1}{4}$	$\frac{3}{16}$		
14	Drawer fronts..............	2	12	4	$\frac{13}{16}$		
15	Drawer front..............	1	27	4	$\frac{13}{16}$	Maple or birch	
16	Drawer front..............	1	27	7	$\frac{13}{16}$		
17	Drawer front..............	1	27	8	$\frac{13}{16}$		
18	Drawer sides..............	6	$17\frac{1}{4}$	$3\frac{7}{8}$	$\frac{1}{2}$		
19	Drawer sides..............	2	$17\frac{1}{4}$	$6\frac{7}{8}$	$\frac{1}{2}$		
20	Drawer sides..............	2	$17\frac{1}{4}$	$7\frac{7}{8}$	$\frac{1}{2}$		
21	Drawer backs.............	2	$11\frac{3}{4}$	$3\frac{7}{8}$	$\frac{1}{2}$	Hardwood	
22	Drawer back..............	1	$26\frac{3}{4}$	$3\frac{7}{8}$	$\frac{1}{2}$		
23	Drawer back..............	1	$26\frac{3}{4}$	$6\frac{7}{8}$	$\frac{1}{2}$		
24	Drawer back..............	1	$26\frac{3}{4}$	$7\frac{7}{8}$	$\frac{1}{2}$		
25	Drawer bottoms............	2	$11\frac{1}{8}$	$16\frac{7}{8}$	$\frac{3}{16}$	3-ply hardwood	
26	Drawer bottoms............	3	$26\frac{1}{4}$	$16\frac{7}{8}$	$\frac{3}{16}$		
27	Drawer slides.............	5	$16\frac{3}{4}$	$2\frac{1}{4}$	$\frac{7}{16}$	Hardwood	
28	Drawer guides.............	5	$16\frac{1}{2}$	1	$\frac{9}{16}$		
29	Door stiles................	4	$15\frac{3}{4}$	$1\frac{1}{2}$	$\frac{13}{16}$		
30	Door rails.................	4	9	$1\frac{3}{4}$	$\frac{13}{16}$	Maple or birch	
31	Door panels...............	2	$13\frac{1}{8}$	$9\frac{1}{2}$	$\frac{3}{8}$		
32	Compartment bottoms.......	2	$17\frac{1}{8}$	$12\frac{7}{8}$	$\frac{3}{16}$	3-ply hardwood	
33	Shelves....................	2	$12\frac{3}{4}$	$16\frac{1}{4}$	$\frac{13}{16}$		
34	Drawer partitions..........	4	$11\frac{3}{8}$	$2\frac{3}{4}$	$\frac{1}{4}$	Hardwood	
35	Drawer partition...........	1	$26\frac{1}{4}$	$2\frac{3}{4}$	$\frac{3}{8}$		
36	Back panel................	1	$53\frac{1}{4}$	23	$\frac{3}{16}$	3-ply hardwood	
37	Shelf supports.............	4				Knape & vogt	Or equal.
38	Shelf pins.................	8				No. 255.	Or equal.
39	Hinges....................	4	2	2		Brass	
40	Locks.....................	2					
41	Knobs....................	8	$1\frac{5}{8}$	$1\frac{5}{8}$	$\frac{15}{16}$	Maple or birch	Turned.

*See figure 21.

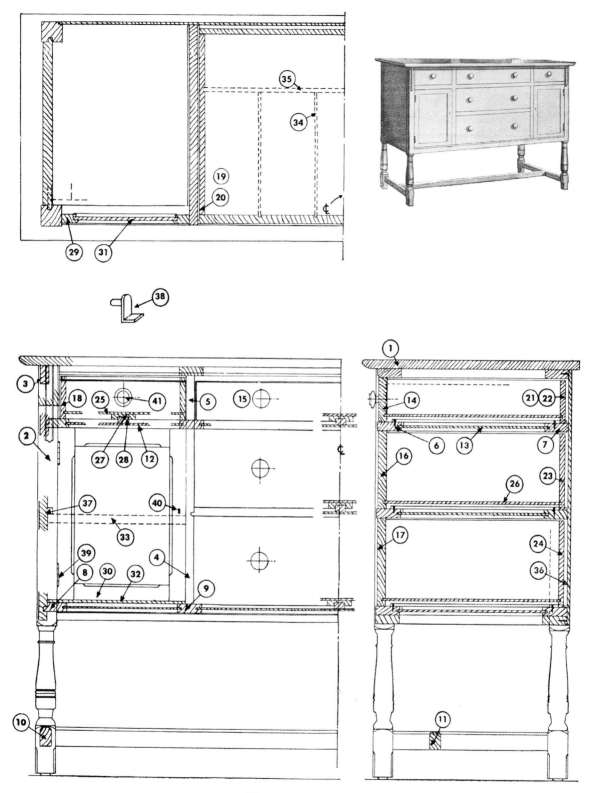

Figure 21. Sideboard.

TABLE XXII. *Bill of materials for china cabinet*

Key*	PART	No.	Length	Width	Thickness	Material	Remarks
			Dimensions (in.)				
1	Top panel	1	36	13	$\frac{13}{16}$	} Maple or birch	
2	Bottom panel	1	36	$12\frac{13}{16}$	$\frac{13}{16}$		
3	Shelves	2	36	$11\frac{9}{16}$	$\frac{13}{16}$		
4	Back piece	1	36	$12\frac{3}{8}$	$\frac{13}{16}$		
5	Front rail	1	36	$1\frac{5}{8}$	$1\frac{5}{16}$		
6	Back panel	1	36	$31\frac{1}{2}$	$\frac{3}{16}$	3-ply hardwood	
7	Side panels	2	44	13	$\frac{13}{16}$	} Maple or birch	
8	Door stiles	4	29	$1\frac{3}{4}$	$\frac{13}{16}$		
9	Door rails	2	$17\frac{3}{4}$	$3\frac{3}{16}$	$\frac{13}{16}$		
10	Door rails	2	$17\frac{3}{4}$	$1\frac{3}{4}$	$\frac{13}{16}$		
11	Glass molds	2	84	$\frac{7}{16}$	$\frac{3}{16}$		
12	Front cornice	1	39	$2\frac{1}{8}$	1		
13	Side cornices	2	14	$2\frac{1}{8}$	1		
14	Glass panes	2	$26\frac{3}{8}$	$15\frac{1}{8}$	$\frac{1}{8}$	Glass	
15	Lock	1				Brass	
16	Hinges	4	$1\frac{3}{4}$	$1\frac{3}{4}$		} Maple or birch	
17	Counter panel	1	40	21	$\frac{13}{16}$		
18	Shelf	1	36	$18\frac{3}{8}$	$\frac{13}{16}$		
19	Bottom panel	1	$35\frac{1}{4}$	$18\frac{7}{8}$	$\frac{3}{8}$	3-ply hardwood	
20	Front rails	2	36	$2\frac{1}{4}$	$\frac{13}{16}$	Maple or birch	
21	Back rails	2	36	$2\frac{1}{4}$	$\frac{13}{16}$	Hardwood	
22	End rails	4	16	$2\frac{1}{4}$	$\frac{13}{16}$		
23	Side panels	2	$27\frac{5}{8}$	20	$\frac{13}{16}$	} Maple or birch	
24	Front base	1	38	$5\frac{3}{8}$	$1\frac{1}{16}$		
25	Side bases	2	$20\frac{5}{8}$	$5\frac{3}{8}$	$1\frac{1}{16}$		
26	Back brace	1	$35\frac{3}{4}$	3	$\frac{13}{16}$	Hardwood	
27	Door stiles	4	26	2	$\frac{13}{16}$	} Maple or birch	
28	Door rails	2	14	$2\frac{1}{8}$	$\frac{13}{16}$		
29	Door rails	2	14	$2\frac{1}{4}$	$\frac{13}{16}$		
30	Door panels	2	$22\frac{3}{8}$	$14\frac{3}{4}$	$\frac{5}{8}$		
31	Panel molds	2	74	$\frac{5}{16}$	$\frac{1}{4}$		
32	Back panel	1	$27\frac{5}{8}$	36	$\frac{3}{16}$	3-ply hardwood	
33	Knobs	2	$1\frac{1}{4}$	$1\frac{1}{4}$	$\frac{3}{4}$	Maple or birch	Turned.
34	Friction catch	1					
35	Pair hinges	2	2	$1\frac{3}{4}$		Brass	

*See figure 22.

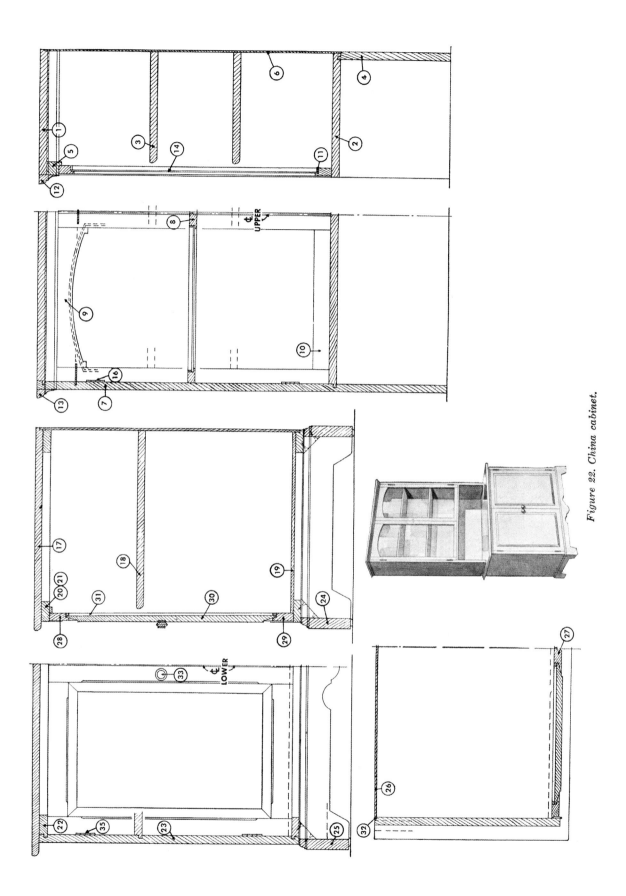

Figure 22. China cabinet.

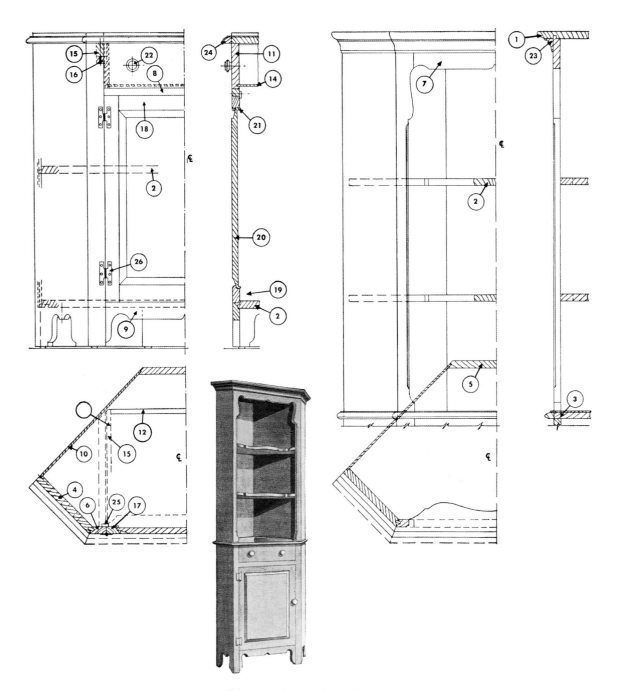

Figure 23. Corner china cabinet.

TABLE XXIII. *Bill of materials for corner china cabinet*

Key*	PART	No.	Dimensions (in.) Length	Width	Thickness	Material	Remarks
1	Top panel	1	37	20	$\frac{13}{16}$	⎫	
2	Shelves	4	34	$17\frac{3}{8}$	$\frac{13}{16}$	⎪	
3	Counter panel	1	34	$17\frac{7}{8}$	$\frac{13}{16}$	⎪	
4	Sides	2	$76\frac{1}{4}$	$8\frac{1}{2}$	$\frac{13}{16}$	⎪	
5	Back piece	1	$76\frac{1}{4}$	11	$\frac{13}{16}$	⎬ Maple or birch	
6	Stiles	2	$76\frac{1}{4}$	2	$\frac{13}{16}$	⎪	
7	Top rail	1	$18\frac{3}{8}$	$3\frac{5}{8}$	$\frac{13}{16}$	⎪	
8	Traverse	1	$18\frac{3}{8}$	$\frac{13}{16}$	$\frac{13}{16}$	⎪	
9	Bottom rail	1	$18\frac{3}{8}$	$4\frac{1}{2}$	$\frac{13}{16}$	⎭	
10	Back panels	2	$71\frac{1}{4}$	$16\frac{1}{2}$	$\frac{3}{16}$	3-ply maple or birch	
11	Drawer front	1	$18\frac{1}{2}$	5	$\frac{13}{16}$	Maple or birch	
12	Drawer back	1	$18\frac{3}{8}$	$4\frac{7}{8}$	$\frac{1}{2}$	⎱ Hardwood	
13	Drawer sides	2	$13\frac{1}{2}$	$4\frac{7}{8}$	$\frac{1}{2}$	⎰	
14	Drawer bottom	1	13	$15\frac{7}{8}$	$\frac{3}{16}$	3-ply hardwood	
15	Drawer cleats	2	$12\frac{5}{8}$	$1\frac{1}{2}$	$\frac{13}{16}$	⎱ Hardwood	
16	Drawer slides	2	$12\frac{5}{8}$	$1\frac{3}{16}$	$\frac{13}{16}$	⎰	
17	Door stiles	2	$22\frac{3}{4}$	$1\frac{5}{8}$	$\frac{13}{16}$	⎫	
18	Door rail	1	$15\frac{1}{4}$	$1\frac{5}{8}$	$\frac{13}{16}$	⎪	
19	Door rail	1	$15\frac{1}{4}$	$1\frac{7}{8}$	$\frac{13}{16}$	⎪	
20	Door panel	1	$19\frac{7}{8}$	16	$\frac{5}{8}$	⎬ Maple or birch	
21	Panel mold	1	75	$\frac{5}{16}$	$\frac{3}{16}$	⎪	
22	Knobs	3	$1\frac{1}{2}$	$1\frac{1}{2}$	$\frac{13}{16}$	⎪	Turned
23	Cornice	1	48	$1\frac{7}{16}$	$\frac{15}{16}$	⎪	
24	Counter mold	1	48	$\frac{1}{8}$	$\frac{13}{16}$	⎭	
25	Door stop	1	58	$\frac{3}{4}$	$\frac{1}{4}$	Hardwood	
26	Pair "H" hinges	1	$2\frac{3}{8}$	$1\frac{3}{8}$		Keeler G—832	Or equal

*See figure 23.

TABLE XXIV. *Bill of materials for single and double bed—Jenny Lind style*

Key*	PART	No.	Dimensions (in.) Length	Width	Thickness	Material	Remarks
1A	Posts—head	2	40	2	2	⎫	
2A	Posts—foot	2	$30\frac{1}{4}$	2	2	⎪	
3A	Top rails—head and foot	2	$39\frac{1}{2}$	$4\frac{1}{2}$	$1\frac{1}{4}$	⎪	
4A	Intermediate rail	1	$39\frac{1}{2}$	$2\frac{1}{4}$	$1\frac{1}{4}$	⎬ Maple or birch	
5A	Bottom rails—head and foot	2	$39\frac{1}{2}$	7	$1\frac{3}{8}$	⎪	
6A	Side rails	2	76	$4\frac{1}{2}$	$1\frac{3}{8}$	⎪	
7A	Spindles	14	14	$1\frac{3}{8}$	$1\frac{5}{8}$	⎭	Turned
8A	Bed rail holders					Stanley No. 1919	Or equal
1B	Posts—head	2	40	2	2	⎫	
2B	Posts—foot	2	$30\frac{1}{4}$	2	2	⎪	
3B	Top rails—head and foot	2	$54\frac{1}{2}$	$4\frac{1}{2}$	$1\frac{1}{4}$	⎪	
4B	Intermediate rail	1	$54\frac{1}{2}$	$2\frac{1}{4}$	$1\frac{1}{4}$	⎬ Maple or birch	
5B	Bottom rails—head and foot	2	$54\frac{1}{2}$	7	$1\frac{3}{8}$	⎪	
6B	Side rails	2	76	$4\frac{1}{2}$	$1\frac{3}{8}$	⎪	
7B	Spindles	18	14	$1\frac{3}{8}$	$1\frac{3}{8}$	⎭	Turned
8B	Bed rail holders	2				Stanley No. 1919	Or equal

*See figure 24.

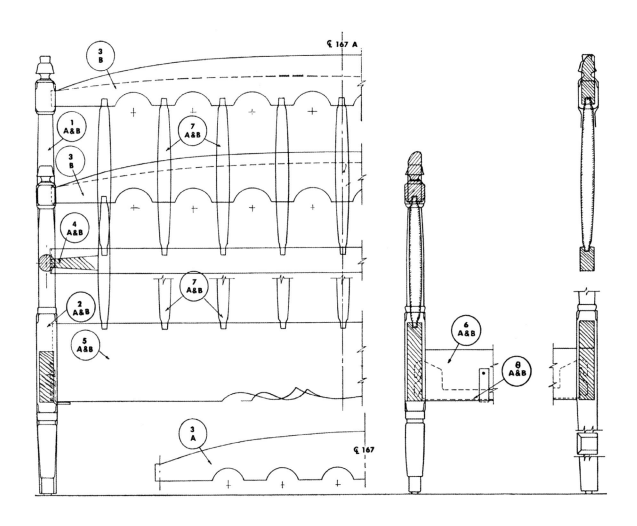

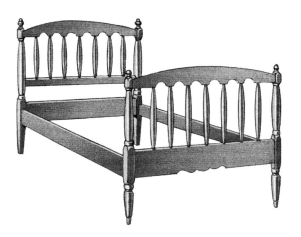

Figure 24. Single and double bed, Jenny Lind style.

TABLE XXV. *Bill of materials for single and double bed—solid head and foot boards*

Key*	PART	No.	Dimensions (in.) Length	Width	Thickness	Material	Remarks
1	Posts—head.................	2	39	2¹⁄₁₆	2¹⁄₁₆		Turned
2	Posts—foot.................	2	30	2¹⁄₁₆	2¹⁄₁₆		Turned
3	Panel—head.................	1	39	24½	⅝	Maple or birch	
4	Panel—foot.................	1	39	15½	⅝		
5	Rails—head and foot........	2	38	4½	1¼		
6	Side rails..................	2	76	4½	1¼		
7	Bed rail holders.............	2				Stanley No. 1919	Or equal
1	Posts—head.................	2	39	2¹⁄₁₆	2¹⁄₁₆		Turned
2	Posts—foot.................	2	30	2¹⁄₁₆	2¹⁄₁₆		Turned
3A	Panel—head.................	1	54¼	24⅝	⅝	Maple or birch	
4A	Panel—foot.................	1	54¼	15¾	⅝		
5A	Rails—head and foot........	2	53	4½	1¼		
6	Side rails..................	2	76	4½	1¼		
7	Bed rail holders.............	2				Stanley No. 1919	Or equal

*See figure 25.

TABLE XXVI. *Bill of materials for chest—six-drawer*

Key*	PART	No.	Dimensions (in.) Length	Width	Thickness	Material	Remarks
1	Top panel..................	1	35¾	20½	¹³⁄₁₆		
2	End panels.................	2	38⅞	19¼	¹³⁄₁₆	Maple or birch	
3	Front rails..................	6	33⅝	2¼	¹³⁄₁₆		
4	Back rails..................	6	33⅝	2¼	¹³⁄₁₆	Hardwood	
5	End rails..................	13	15⁵⁄₁₆	2¼	¹³⁄₁₆		
6	Dust bottoms..............	5	29¾	15⁵⁄₁₆	³⁄₁₆	3-ply hardwood	
7	Drawer stile..............	1	4	2½	1⅛		
8	Drawer fronts..............	2	16¼	4³⁄₁₆	¹³⁄₁₆		
9	Drawer front..............	1	33¼	6³⁄₁₆	¹³⁄₁₆	Maple or birch	
10	Drawer front..............	1	33¼	7³⁄₁₆	¹³⁄₁₆		
11	Drawer front..............	1	33¼	8³⁄₁₆	¹³⁄₁₆		
12	Drawer front..............	1	33¼	9³⁄₁₆	¹³⁄₁₆		
13	Drawer sides..............	4	18¾	3⅞	½		
14	Drawer sides..............	2	18¾	5⅞	½		
15	Drawer sides..............	2	18¾	6⅞	½		
16	Drawer sides..............	2	18¾	8⅞	½		
17	Drawer sides..............	2	18¾	9⅞	½	Hardwood	
18	Drawer backs..............	2	15⅝	3⅞	½		
19	Drawer back..............	1	32⅝	5⅞	½		
20	Drawer back..............	1	32⅝	6⅞	½		
21	Drawer back..............	1	32⅝	7⅞	½		
22	Drawer back..............	1	32⅝	8⅞	½		
23	Drawer bottoms............	2	15⅛	18¼	³⁄₁₆	3-ply hardwood	
24	Drawer bottoms............	4	32⅛	18¼	³⁄₁₆		
25	Drawer slides.............	6	18¼	2¼	⁷⁄₁₆	Hardwood	
26	Drawer guides.............	6	18	1	⁹⁄₁₆		
27	Back panel................	1	38⅞	33⅝	³⁄₁₆	3-ply hardwood	
28	Front base................	1	35¾	5⅝	1¹⁄₁₆	Maple or birch	
29	End bases................	2	19⅞	5⅝	1¹⁄₁₆		
30	Back brace................	1	33⅝	3½	¹³⁄₁₆	Hardwood	
31	Knobs....................	2	1¾	1¾	1	Maple or birch	Turned
32	Drawer pulls...............	8	5½	1	1⁵⁄₁₆		Turned

*See figure 26.

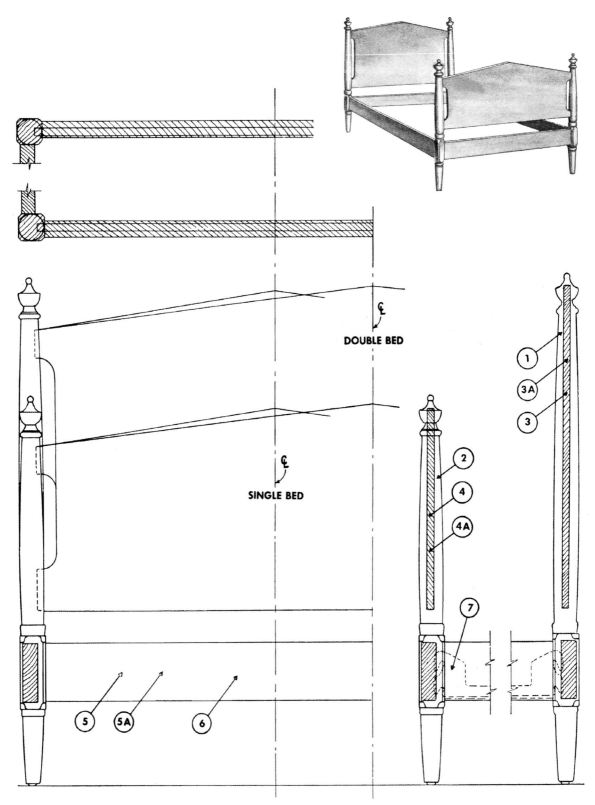

DOUBLE BED

SINGLE BED

Figure 25. Single and double bed, solid head and footboards.

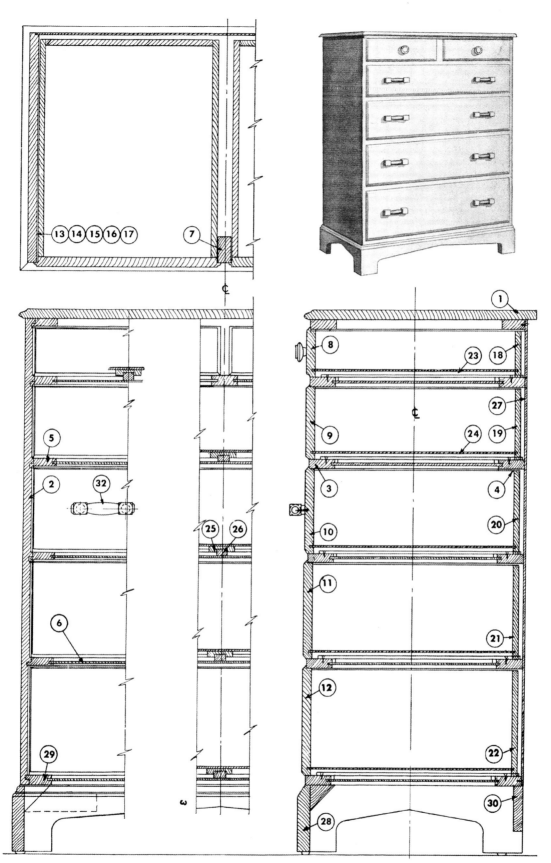

Figure 26. Chest—six-drawer.

TABLE XXVII. Bill of materials for dresser—four-drawer

Key*	PART	No.	Dimensions (in.)			Material	Remarks
			Length	Width	Thickness		
1	Top panel	1	41	21	13/16		
2	End panels	2	29 1/4	20	13/16	Maple or birch	
3	Front rails	5	39	2 1/4	13/16		
4	Back rails	5	39	2 1/4	13/16	Hardwood	
5	End rails	10	15 3/4	2 1/4	13/16		
6	Dust bottoms	4	34 3/4	15 3/4	3/16	3-ply hardwood	
7	Drawer front	1	38 1/2	4 1/4	13/16		
8	Drawer front	1	38 1/2	6 1/4	13/16	Maple or birch	
9	Drawer front	1	38 1/2	7 1/4	13/16		
10	Drawer front	1	38 1/2	8 1/4	13/16		
11	Drawer sides	2	19 1/4	3 7/8	1/2		
12	Drawer sides	2	19 1/4	5 7/8	1/2		
13	Drawer sides	2	19 1/4	6 7/8	1/2		
14	Drawer sides	2	19 1/4	7 7/8	1/2	Hardwood	
15	Drawer back	1	38	3 7/8	1/2		
16	Drawer back	1	38	5 7/8	1/2		
17	Drawer back	1	38	6 7/8	1/2		
18	Drawer back	1	38	7 7/8	1/2		
19	Drawer bottoms	4	37 1/4	18 3/4	3/16	3-ply hardwood	
20	Drawer slides	4	18 1/4	2 1/4	7/16	Hardwood	
21	Drawer guides	4	18 1/2	1	9/16		
22	Back panel	1	39	29	3/16	3-ply hardwood	
23	Drawer pulls	8	5 1/2	1 3/8	1		Turned
24	Front base	1	41	5 3/8	1 1/16	Maple or birch	
25	End bases	2	20 1/2	5 3/8	1 1/16		
26	Back brace	1	39	2 3/4	13/16	Hardwood	

*See figure 27.

TABLE XXVIII. Bill of materials for dressing table

Key*	PART	No.	Dimensions (in.)			Material	Remarks
			Length	Width	Thickness		
1	Top panel	1	36	20	13/16		
2	Legs	4	29 1/4	1 5/8	1 5/8		Turned
3	Aprons	2	15	5 5/8	13/16	Maple or birch	
4	Apron	1	28 1/2	5 5/8	13/16		
5	Rails—front and back	4	30	2 1/4	13/16		
6	Rails—ends	4	13 1/8	2 1/4	13/16	Hardwood	
7	Dust bottom	1	3 1/8	26 1/4	3/16	3-ply hardwood	
8	Drawer front	1	27 3/4	4	13/16	Maple or birch	
9	Drawer back	1	27 1/2	3 7/8	1/2	Hardwood	
10	Drawer sides	2	15 3/4	3 7/8	1/2		
11	Drawer bottom	1	15 3/8	27	3/16	3-ply hardwood	
12	End stretchers	2	14 1/4	1 1/2	3/4	Maple or birch	
13	Back stretcher	1	27 3/4	1 1/2	3/4		
14	Center slide	1	15 1/4	2 1/4	7/16	Hardwood	
15	Center guide	1	15	1	9/16		
16	Knobs	2	1 9/16	1 9/16	13/16	Maple or birch	Turned

*See figure 28.

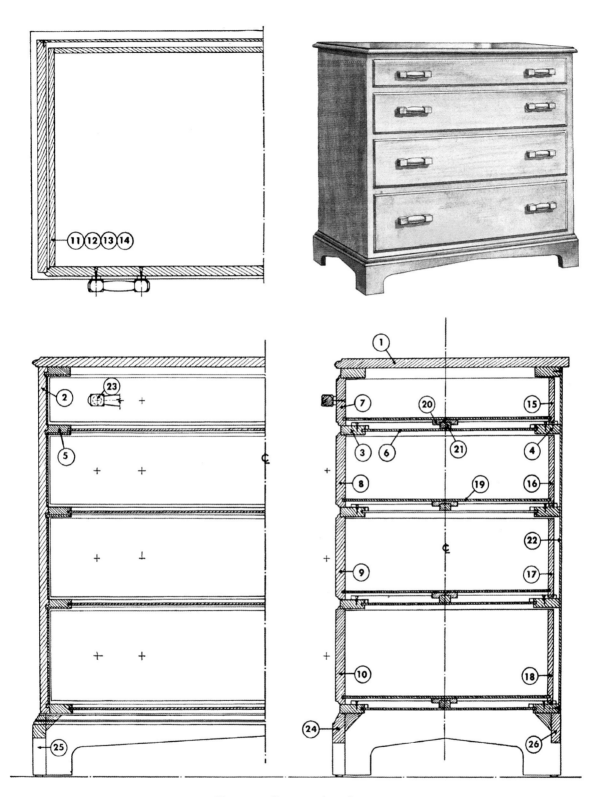

Figure 27. Dresser—four-drawer.

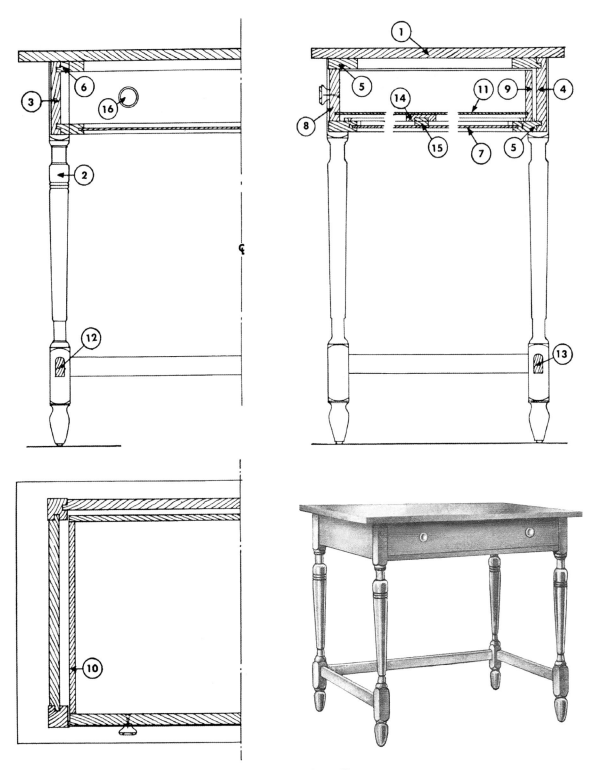

Figure 28. Dressing table.

10. Joints and Fastenings

Figures 29 to 35, inclusive, show common types of joints.

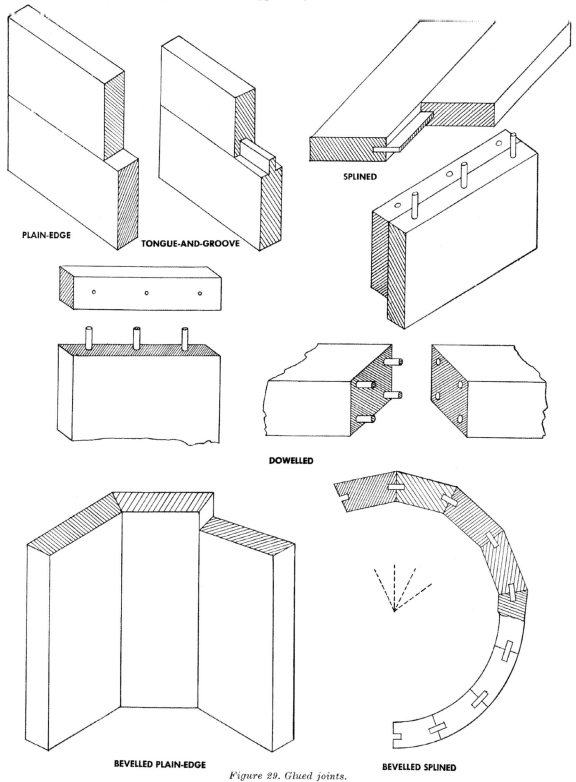

PLAIN-EDGE

TONGUE-AND-GROOVE

SPLINED

DOWELLED

BEVELLED PLAIN-EDGE

BEVELLED SPLINED

Figure 29. Glued joints.

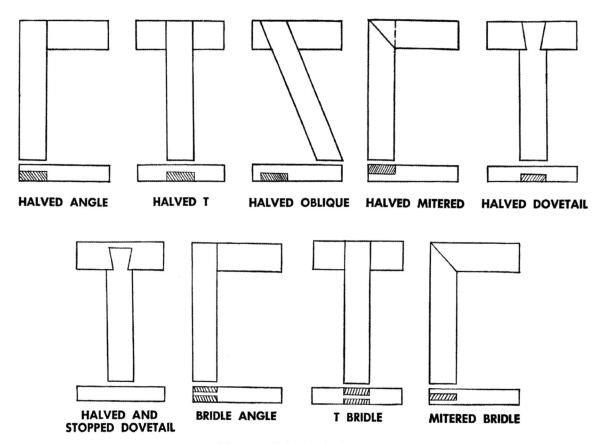

HALVED ANGLE **HALVED T** **HALVED OBLIQUE** **HALVED MITERED** **HALVED DOVETAIL**

HALVED AND STOPPED DOVETAIL **BRIDLE ANGLE** **T BRIDLE** **MITERED BRIDLE**

Figure 30. Halved or bridle joints.

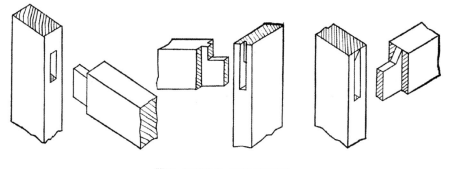

STUB MORTISE AND TENON

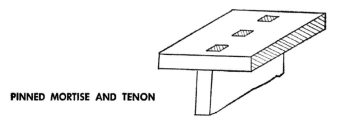

PINNED MORTISE AND TENON

Figure 31. Mortise and tenon joints.

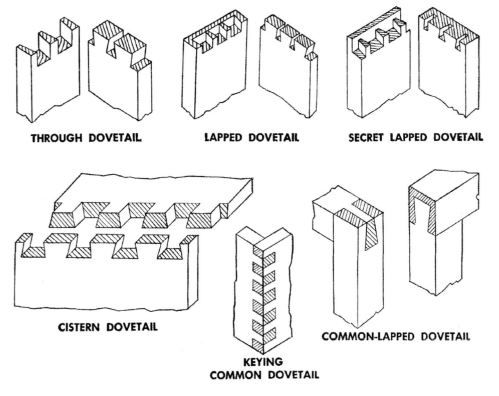

THROUGH DOVETAIL　　**LAPPED DOVETAIL**　　**SECRET LAPPED DOVETAIL**

CISTERN DOVETAIL

KEYING COMMON DOVETAIL

COMMON-LAPPED DOVETAIL

Figure 32. Dovetail joints.

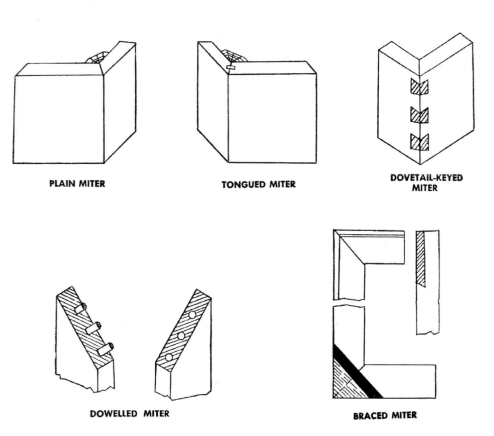

PLAIN MITER　　**TONGUED MITER**　　**DOVETAIL-KEYED MITER**

DOWELLED MITER　　**BRACED MITER**

Figure 33. Mitered joints.

51

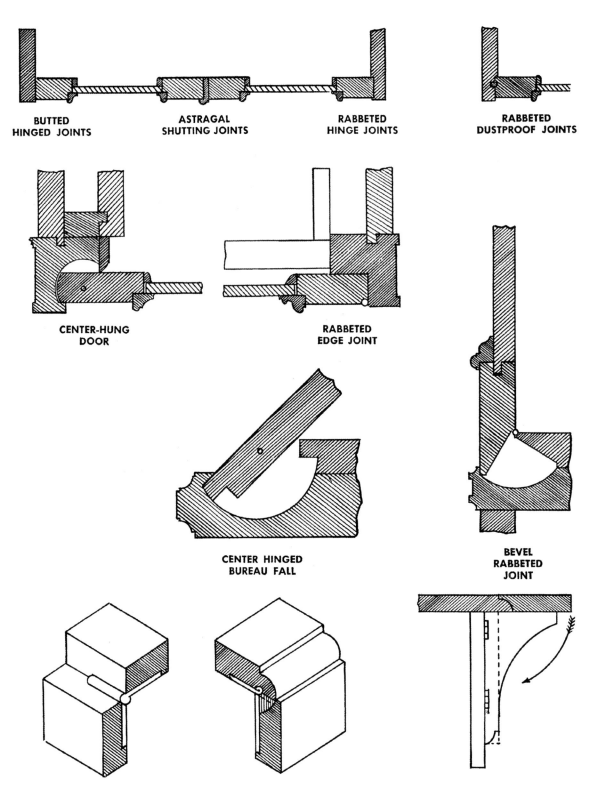

BUTTED HINGED JOINTS

ASTRAGAL SHUTTING JOINTS

RABBETED HINGE JOINTS

RABBETED DUSTPROOF JOINTS

CENTER-HUNG DOOR

RABBETED EDGE JOINT

CENTER HINGED BUREAU FALL

BEVEL RABBETED JOINT

PLAIN HINGED RULE JOINT HINGED BRACKET

Figure 34. Hinged and shutting joints.

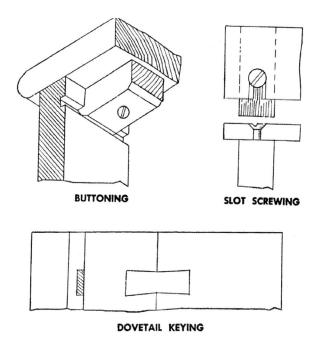

BUTTONING

SLOT SCREWING

DOVETAIL KEYING

Figure 35. Fastenings.

CHAPTER 4

WOODWORKING MACHINERY

Section I. INTRODUCTION

11. General

Wood can be processed rapidly and accurately by machine. Years of practice are required for skilled handwork, but the operation of a machine often can be learned in a few weeks. The following paragraphs describe operations on some of the most frequently used machines. No machine should be operated except by personnel qualified either by instruction or experience.

Section II. CIRCULAR SAW BENCHES

12. Types

Many types of circular saw benches are used in woodworking shops. Principal types are described below.

a. CUT-OFF SAWS. Cut-off saws, used only for crosscutting, may be the overhead swing type for attaching to ceiling, post, or wall; or the self-contained pedestal type which is bolted to the floor.

b. RIPSAWS. Ripsaws normally are used only for ripping. There are three general types: hand-feed, roll-feed, and chain-feed or straight-line ripsaws. The last is used for edging and jointing as well as for ripping.

c. UNIVERSAL DOUBLE-ARBOR SAWS. Universal double-arbor saws have a double arbor mounting both ripsaw and crosscut-saw blades, and can therefore be used both for ripping and crosscutting. They are ideal for shops which do both ripping and crosscutting but do not have enough space or production for separate ripsaws and crosscut saws.

d. SINGLE-ARBOR VARIETY SAWS. Single-arbor variety saws are better than other circular saw benches for post engineer purposes, because they can do a wide variety of work, including ripping, crosscutting, mitering, beveling, grooving, tenoning, dadoing, and cutting moldings. A number of saw blades and attachments are required for these operations. Some common types of variety saws and attachments are:

(1) Tilt-table type. (See fig. 36.)

(2) Tilt-arbor type. (See fig. 37.)

(3) Tilt-table type with removable throat plate for dado heads.

(4) Tilt-table type with sliding table section for exceptionally wide cutoffs and lateral adjustment for dado heads. (See fig. 38.)

(5) Tilt-table type with mortising and boring attachment.

13. Saw Blades

The efficiency of a saw table depends mainly on the type of saw used and the condition of its blade. Three types of saw blades are used on circular saw benches. (See fig. 39.)

a. CIRCULAR RIPSAW. The circular ripsaw (fig. 39 ①) is best for general ripping of miscellaneous woods. The shape of the tooth, the gullet space, and the clearance on the back of the tooth are important factors. The standard shape has a series of chisel teeth with plenty of back clearance to prevent drag. Tests have shown that a tooth space of $1\frac{1}{4}$ inches, point to point, is ideal for lumber up to 2 inches thick. For coarse ripping on heavy work, wider spacing is desirable. For fine ripping, closer spacing gives better results.

b. CIRCULAR CROSSCUT SAW. The circular crosscut saw (fig. 39②) is best for general crosscutting. The spacing on cut-off saws in sizes from 12 to 16 inches ranges from $\frac{7}{16}$ to $\frac{7}{8}$ inch from point to point, depending on whether the saw is used for fine or coarse work. For general crosscutting, a $\frac{1}{4}$-inch saw with 100 teeth is recommended.

Figure 36. Tilt-table saw bench with table tilted.

c. MITER SAW. The hollow-ground miter saw (fig. 39 ③) is also called a combination saw or planer saw. It has two types of teeth; cutting teeth shaped like those in a cross-cut saw, which do the rough cutting; and chisel-shaped raker or cleaner teeth which give the kerf a smooth finish. Raker teeth are slightly shorter than cutting teeth, and have a wider throat for cleaning out chips. Usually, every fifth to eighth tooth is a raker tooth. Miter saws usual-

Figure 37. Tilt-arbor saw bench with arbor tilted.

ly are employed on single-arbor variety saw benches for both ripping and crosscutting. Since the saw is hollow-ground, teeth need not be set. However, clearance is limited, so the saw must not be crowded.

14. Safety Measures

The following safety rules must be observed when using circular saws:

a. Use a guard on the saw unless the use of a jig is approved by the supervisor.

b. Do not remove the splitter when using the saw for ripping. It is one of the most important guards on the saw bench.

> NOTE. Guards and splitters have been removed in many illustrations in this manual to give a clearer picture of the cutting action of the machines. This must not be construed to mean that this practice is acceptable.

c. Adjust saw so the blade projects ⅛ inch above the stock.

d. Do not stand directly behind or in line with the saw when it is in operation; a piece thrown back as a result of binding can cause serious injury.

e. Do not reach over a revolving saw. Have an assistant guide stock from the back of the table.

f. Never wear loose clothing. Roll sleeves up above elbows and tuck necktie securely inside shirt.

g. Use a push stick (fig. 40) between fence and saw for all narrow ripping.

h. Avoid any freehand ripping or crosscutting. Use fence or cut-off gauge.

i. Before ripping stock, make sure at least one edge is straight and free from projections. Hold this side against the fence. If possible, joint all stock before ripping.

j. Attach a clearance block to the front face

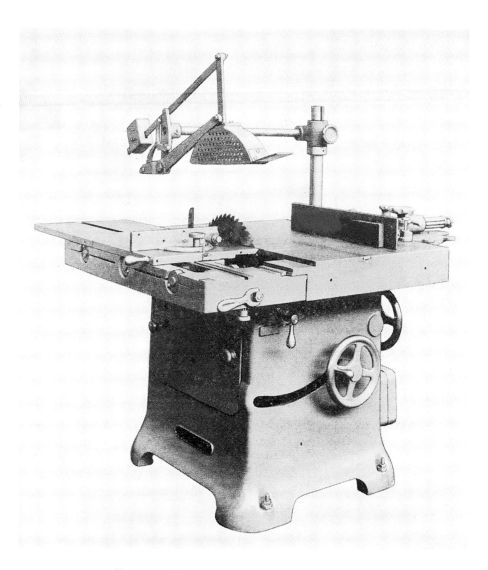

Figure 38. Tilt-arbor saw bench with sliding table.

① *Ripsaw.* ② *Crosscut saw.* ③ *Miter saw.*
Figure 39. Types of circular saw blades.

of the ripping fence whenever it is used as a stop in crosscutting narrow stock.

k. Open switch or remove circuit fuses before changing saw blade or making repairs or adjustments on the machine.

l. Do not use a dull saw or one that does not have the proper set.

m. Always clean off saw tables with a stick or brush; never use the hands.

in position. To accommodate wide stock, move fence carriage away from the saw to a second or third set of locating holes. Tapered pins hold it in position.

(4) See that table is level and at right angles to the saw and that guard is in position over the saw.

(5) Standing at left of the work, place straight or jointed edge of the stock against

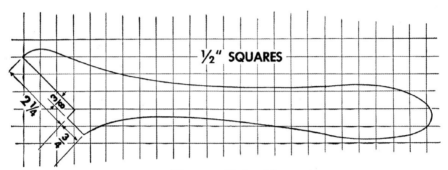

Figure 40. Push stick

n. Keep machine and floor free of dirt and all debris. If floor is slippery, provide mats.

o. Never look around while operating the machine.

p. Before starting the saw, make certain that arbor nut is properly tightened and that guard is in place.

15. Using Ripsaw

a. SIMPLE RIPPING. The easiest operation at the circular saw is simple ripping:

(1) Attach a ripsaw on the arbor. If a universal saw bench is used, bring ripsaw into position.

(2) On a plain-table saw bench, raise or lower the saw table or saw as necessary and set the clamp so the table or arbor cannot move.

(3) Using the scale on the table, set the ripping fence and lock it. The width set should be from $\frac{1}{16}$ to $\frac{1}{8}$ inch greater than the desired width of the finished stock. Extra width permits finishing the ripped edge. For accurate settings, use micrometer adjustment on the fence. Make sure fence carriage is resting in locating holes and is securely fastened to the table. See that thumbscrews which hold the fence in position are set and that fence face is at right angles to the table surface. Lock fence

the fence and feed stock against the saw. (See fig. 41.)

(6) Use a push stick when ripping stock into narrow strips. (See fig. 42.) Cut a notch in the end of the push stick at such an angle that the handle extends well above the fence; make sure notch is deep enough to hold stock securely on the table. Because guard is in the way when narrow strips are ripped, set saw to project $\frac{1}{8}$ inch above the stock.

(7) When ripping wide stock, support stock with the hands and hold it against saw with the body. Keep to the *right* side of the machine and force the stock steadily ahead, holding it firmly against ripping fence. (See fig. 43.)

(8) Have an assistant tail the machine if possible.

b. RABBETING. Cut rabbets by making two simple ripping cuts at right angles to each other, as follows:

(1) Set ripping fence for desired width of face kerf and adjust the saw vertically for depth of cut.

(2) With edge of stock against the fence, make the first cut.

(3) Reset saw and fence.

(4) With stock resting on edge and a face

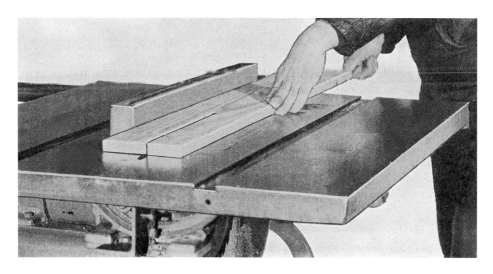

Figure 41. Ripping average stock.

against the fence, make second cut to remove the waste strip and form rabbetted edge.

(5) To form a tongue, rabbet both sides on the same edge of a board.

c. GROOVING. Make grooved joints as follows:

(1) Set saw vertically to cut groove to desired depth and adjust fence to locate the groove.

(2) Saw the first kerf. To enlarge groove, move fence ⅛ inch for each cut.

(3) Use a stop when groove is not needed for the entire length of stock.

d. RIPPING AT AN ANGLE. Stock can be chamfered, beveled, or ripped at an angle by tilting either fence or table. If an exact setting of fence is wanted for subsequent cuts, table setting may be adjusted for the angle cut; if the table setting is retained, fence setting may be adjusted.

(1) *With fence tilted.* To rip stock at an angle by tilting the fence, adjust table and saw for simple ripping and tilt fence at angle of desired cut. Narrow stock lies against the fence easily and can be handled in the same way as for simple ripping. With wide stock,

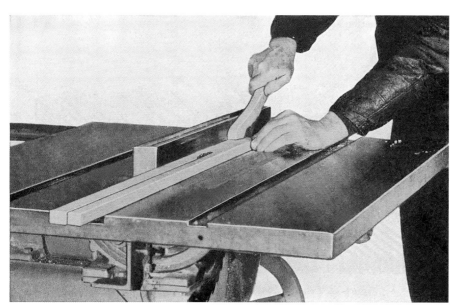

Figure 42. Ripping narrow stock with push stick.

Figure 43. Ripping wide stock.

however, the overhanging weight may pull the stock away from the fence, changing the angle of the cut. Cut wide stock at correct angle by one of the following methods:

(a) Hold wood against the fence with left hand and keep stock from tilting with right hand.

(b) Use a feather board. (See figs. 44 and 66.) The feather board holds stock against the fence and acts as a guard, protecting the operator from the saw; the latter is important because the saw guard cannot be in place for this operation. Raise or lower the feather board as required, by clamping different thicknesses of stock under it.

Note. Cut bevel or chamfer completely around stock with miter or combination saw.

(2) *With table tilted.* To rip stock at an angle by tilting the table, adjust table at desired angle and lock the fence in position at *left* side of table. (See fig. 45.) Slope of table holds stock against the fence and keeps it from crowding saw. Stand at left of table because waste stock, on right, drops against saw and is thrown to front of machine.

16. Using Crosscut Saw

a. SQUARING ENDS. Squaring the ends of stock is one of the main uses of crosscut saws. A miter or universal cut-off gauge is used instead of the ripping fence to guide the cut. To square stock ends—

(1) Move the ripping fence to the right or remove it entirely.

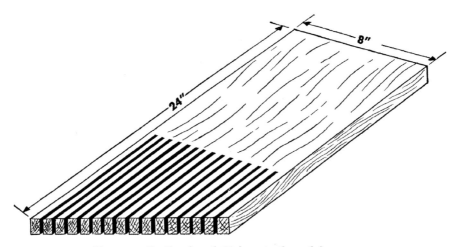

Figure 44. Feather board. Make cuts through lower section of board with band saw. The remaining strips are flexible, acting as spring for holding stock.

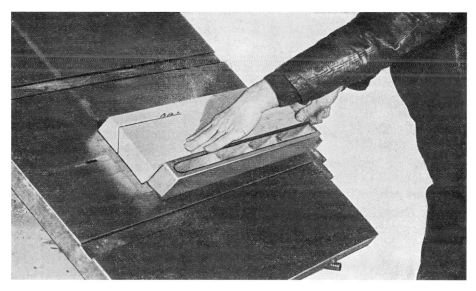

Figure 45. Ripping stock at an angle.

(2) Check the universal gauge for squareness by placing the gauge upside down against the front edge of the table while the slide is loose in miter gauge slot or groove of the saw table. When the gauge is squared, clamp parts securely together, turn gauge over, and replace it in the groove on the left half of the table. Usually, an engraved line on the table shows the exact location of the saw with relation to width of stock to be cut.

(3) Before starting machine, see that all movable parts are securely clamped and that table surface is at right angles to saw. For all simple cut-off sawing, have guard in place over the saw.

(4) Place stock on table with jointed surface down and jointed edge against universal gauge. Hold stock securely so it will not move lengthwise along the gauge. A stop is not used when squaring the first end but may be used for cutting to length.

(5) Push stock against the saw, allowing enough stock to project beyond the line of the saw to square an entire end. (See fig. 46.) Several pieces can be cut at one time.

(6) Before squaring the second end, lay off desired finished length of stock along table, measuring from left cutting edge of saw, and clamp a stop rod in the slot of the miter or universal gauge at that point. If stop rods sup-

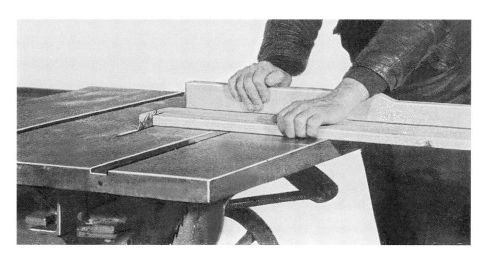

Figure 46. Crosscutting to square first end.

plied with gauge are too short, use strips of wood clamped to miter or cut-off gauge. When handling long stock, remove ripping fence from table top.

(7) Push stock against saw and cut to finished length. (See fig. 47.) When duplicate short-length stock is required and a short rod with pin stop is used, make sure the stop rod does not extend in the path of the saw. Reverse the rod end for end if necessary or use a wooden stop block clamped to the gauge with a C-clamp or small hand screw.

b. CUTTING LONG STOCK. To cut long stock into duplicate shorter pieces—

(1) Square first end of stock as in *a* above.

(2) Attach wooden clearance block to rip-

ping fence with thumbscrew. Adjust fence on right side of table so distance from clearance block to right edge of saw equals desired finished length of stock.

(3) Butt squared end of stock against clearance block and cut to desired length, feeding stock with universal gauge. (See fig. 48.)

c. CUTTING WIDE STOCK. When stock is too wide to be handled conveniently between saw and universal gauge, use a sliding table and cut stock as follows:

(1) Remove the universal gauge. Pin miter cut-off gauge in the locating holes on sliding table and clamp it in place.

(2) Move ripping fence well out of the way. Withdraw locking pin under the left edge

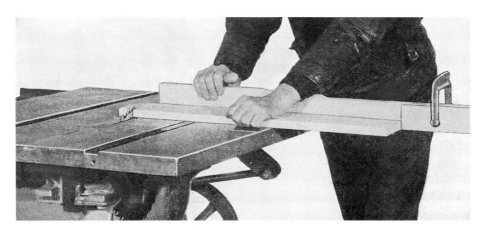

*Figure 47. Trimming board to finished length.
Note use of stop.*

*Figure 48. Crosscutting long piece to short duplicate
lengths, using clearance block on ripping fence.*

of the table, releasing table so it is free to move on rollers.

(3) Move sliding table back, to allow the stock to be placed against the miter gauge, clear of saw teeth. Use a stop to keep stock from creeping.

(4) Hold stock with left hand hooked over far edge. Place right hand on handle of miter gauge. Push table forward and cut off first end.

(5) Reset stop to cut stock to desired length.

(6) Reverse stock, end for end, keeping jointed edge against the gauge and squared end against the stop. Cut off second end.

d. DADOES AND GAINS. To cut dadoes and gains with a crosscut saw—

(1) Adjust table and fence as for simple crosscutting. Set saw blade so amount of saw above table equals depth of the dado or gain.

(2) Butt stock against the fence or against a stop adjusted for each cut. Number of cuts determines width of dado or gain. Move the stock one saw thickness for each cut.

(3) If limits of the dado or gain are laid out with a knife, make one cut just inside each knife line. Remove remaining stock by a series of cuts.

(4) When corresponding dadoes are to be made, as on opposite ends or sides of a cabinet, cut both dadoes with the same set-up.

(5) Cut gain joints or stop dadoes only part way across the stock. Finish ends of the cuts with a sharp chisel.

17. Mitering

a. MITERING EDGE. To cut a simple miter from edge to edge of a piece of stock, as for picture frames—

(1) Adjust table and saw in the same way as for crosscutting. Set miter cut-off gauge or universal gauge at desired angle of cut. When miter cut-off gauge is used with a sliding table to make cuts for square, hexagonal, or octagonal work, the tapered locating holes for the pins place the gauge accurately.

(2) After setting to the desired angle, check the universal or sliding gauges to see that angles do not change when securing the gauge. These gauges are used on either right or left side of the saw.

(3) Attach a stop to the gauge, to keep stock from slipping. (See fig. 49.)

(4) Cut all miters from the same working face or edge to insure accurate fit of joints.

b. MITERING FACE. The saw cannot always be raised enough to cut a miter from face to face of the stock. To miter the face—

(1) Tilt table at angle of desired cut.

(2) Set miter or universal gauge at right angles to plane of saw. If miter gauge is used

Figure 49. Miter cutting with miter gauge set at an angle.

Figure 50. Mitering face of stock.

with sliding table, clamp gauge firmly in position. If universal gauge is used, make sure it is square. (See par. 16a (2).) See that fence is clamped securely, clear of the work.

(3) Clamp stop rod or stop block to the gauge to prevent slipping.

(4) Standing at left of table, push work into saw. (See fig. 50.)

c. COMPOUND MITERS. A compound miter or hopper cut is one in which the cut slants from face to face and from edge to edge, as on a box with slanting sides. To make it, set both gauge and table at an angle. Figure 51 gives settings for table and miter gauge to cut square, hexagonal, or octagonal box sides at any angle up to 45°. To cut a compound miter—

(1) Determine angle of taper needed for box sides.

(2) Locate desired angle of taper on curve for box desired. (See fig. 51.) Read angles of table setting on horizontal scale and gauge setting on vertical scale; see example on chart.

(3) Set table and gauge at these angles, and cut miter. (See fig. 52.)

d. SPLINED MITER JOINTS. Miter joints do not make strong connections between box sides, unless they are reinforced. The insertion of

splines increases their strength considerably. Splined joints can be made with the table tilted, as in figure 53, or the fence tilted, as follows:

(1) Tilt ripping fence so beveled edge of the stock rests on the saw table. Move fence close to the saw so the cut will be close to the inside edge of the bevel, where stock is thicker and cut can be deeper. Use micrometer knob for fine adjustments of fence.

(2) Set the saw for desired depth of cut and clamp it securely.

(3) Use a push stick cut square on the end, or at the proper angle for a compound miter, to keep stock from tilting and to hold bevel on the table. A feather board (par. 15d (1)) helps keep the work at the desired angle against the fence.

(4) Using same settings for fence and saw, cut groove in both bevels of miter joint.

(5) Insert spline, making sure its grain runs across the joint, approximately at right angles to bevel.

18. Using the Dado Head

Although grooves, rabbets, and dadoes can be made with ripsaws and crosscut saws, the dado head is preferred for production work. A dado

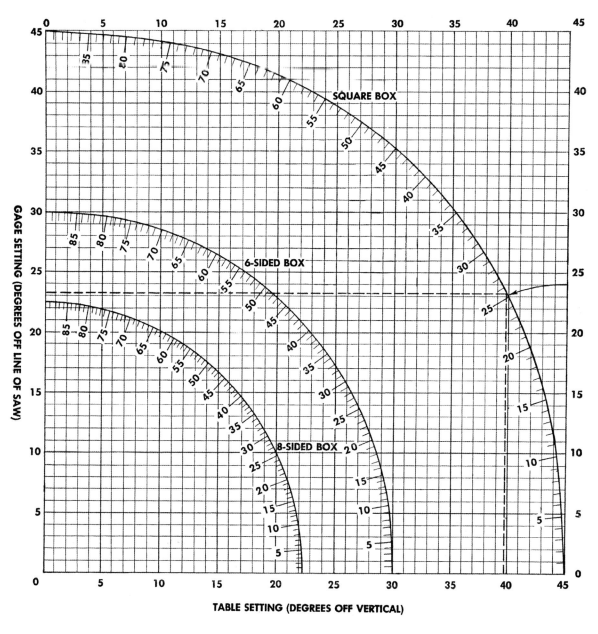

TABLE SETTING (DEGREES OFF VERTICAL)

SQUARE BOX WITH SIDES TAPERED 25°.
SEE GRAPH FOR METHOD OF FINDING GAGE
AND TABLE SETTINGS

23°5'
(GAGE)

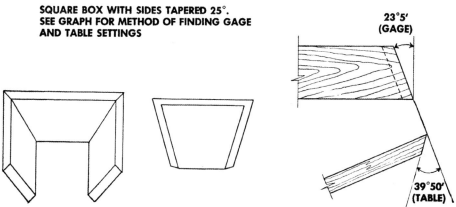

39°50'
(TABLE)

Figure 51. Chart for cutting compound miters.

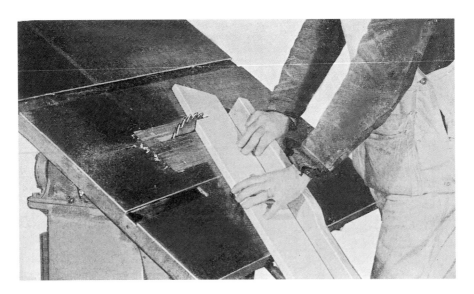

Figure 52. Cutting compound miter.

*Figure 53. Making splined miter joints with
table tilted.*

head can also be used to cut tenons. (See par. 21*d.*) A dado head consists of two outside cutters similar to combination saws which cut along the outside limits of the work, and several chisel-like inside cutters of various widths which remove the stock left between the outside kerfs. Depending on the number of cutters used, a dado head cuts a ⅛-inch to a 2-inch kerf. It cuts equally well with or across the grain. Some saw tables slide sidewise to permit insertion of the wider head. On others it is necessary to use a special plate with opening large enough to accommodate the dado head.

a. SETTING UP DADO HEAD. To set up the saw table for a dado head—

(1) Open table, remove the circular saw, and screw a dado head sleeve on the saw arbor.

(2) Place cutters on the sleeve arbor, making sure cutting edges revolve to the front. Slip filler collars or washers on arbor, outside of blades, and replace the nut. Use one outside cutter for a ⅛-inch cut, both outside cutters for a ¼-inch cut, and as many inside cutters as needed for wider cuts.

(3) Close table by sliding it laterally toward the dado head, or use dado table plate. Allow about 1/16 inch clearance.

b. CUTTING GROOVES AND RABBETS. To cut grooves and rabbets, set the dado head for width of cut and raise or lower the table to cut the desired depth. Adjust ripping fence to locate the groove or rabbet and proceed as for simple ripping. Make sure that stock edges are parallel. If a rabbet is to be cut on a piece without parallel sides, the edge to be rabbeted is held against the fence. In this case, a wooden faceplate must be attached to the fence so that the cutter head touches it and makes a clean cut.

c. CUTTING DADOES AND GAINS. Cut dadoes or gains in the same way as grooves and rabbets, but make cuts across the grain. For right-angle cuts, use miter cut-off gauge with sliding table or a universal gauge checked for squareness. To cut dadoes or gains at an angle, adjust gauges at desired angle.

d. SPECIAL HEADS. Special grooving and molding heads can be used on the saw arbor. These consist of round or square metal heads with specially-shaped knives and spurs attached. The spurs slit the wood ahead of the cutters so sides of the cuts are smooth.

19. Cutting Contours and Moldings

a. SAW-KERF METHOD. Concave or convex curves and moldings can be cut to approximate shape by the following method:

(1) To shape a concave curve, cut stock to finished width and thickness, making opposite edges parallel. Draw a center line on the end of the stock and lay out the contour desired. Adjust saw for the deepest kerf and cut on center line of stock. (See fig. 54①.) Move fence ⅛-inch away from the saw and decrease depth of kerf to follow the contour line. Cut on both sides of the first kerf, holding each edge in turn against the ripping fence. (See fig. 54②.) Continue moving fence and changing depth of cut until the shape is roughed out. (See fig. 54③.) Finish the inside surface with sharp gouges or a core-box plane. Smooth the curved surface with molding scrapers and with sandpaper held over a 4- or 5-inch block of wood shaped to fit the contour of the curve. Use coarse sandpaper first, then finer grades.

(2) In cutting a convex curve the method is the same, except that the center line indicates the shallowest cut. In shaping a semicircular piece in this fashion, the portion of waste stock which would be removed by the last two saw cuts should be finished off by hand, to leave bearing surfaces against fence and saw table and so prevent stock from rolling sidewise.

b. KERF-AND-MITER METHOD. Convex curves and moldings can be shaped by a combination of the kerf method and mitering, as follows:

(1) Cut stock to finished width and thickness. Draw center line on end of stock and lay out the contour desired.

(2) Make square shoulder cuts. (See 1 nd 2, fig. 55.)

(3) Tilt fence so the next cut will be tangent to the contour line. Continue adjusting

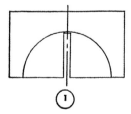 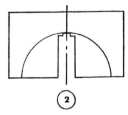 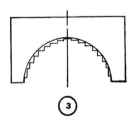

Figure 54. Kerf method of contour cutting.

fence and making tangent cuts until no more stock can be removed by this method (See 3 and 4, fig. 55.)

(4) Use saw-kerf method (*a* above) to cut out remaining stock. (See 5, fig. 55.)

(5) Finish the surface with molding scrapers ground to shape and with sandpaper on blocks formed to fit the contour.

c. OBLIQUE-SAWING METHOD. Concave curves and moldings can also be shaped by pushing the stock across a ripsaw or combination saw at an angle instead of in a line parallel with its plane. Settings for fence and saw are determined by using a parallel rule (fig. 56) or sighting.

(1) *Using parallel rule.* To cut contours using a parallel rule:

(*a*) Draw curve or contour on end of stock.

(*b*) Set saw for greatest depth of cut, the radius R of the curve. (See fig. 57①.)

(*c*) Set parallel rule so the distance between long sides equals diameter D of the curve

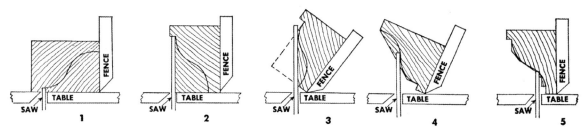

Figure 55. Cutting contours by kerf-and-miter method.

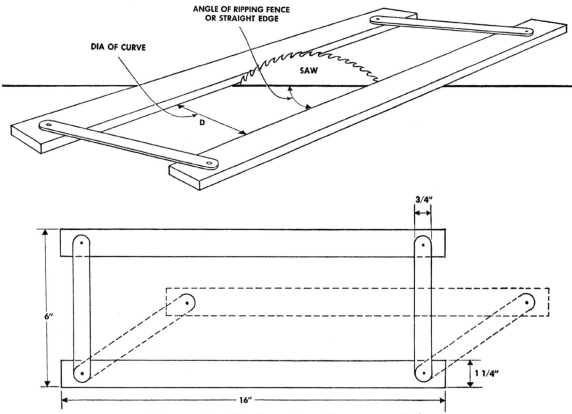

Figure 56. Parallel rule used in cutting contours. Built frame of ⁹⁄₁₆-inch stock. Fasten corners with short flat-head screws countersunk and tightened so the rule is held in any desired position.

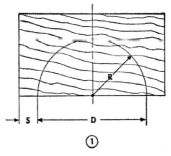 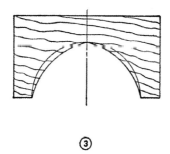 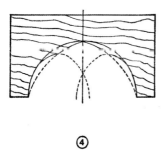

Figure 57. Setting parallel rule for desired contour.

and place rule on saw table with saw teeth just touching inside edges of the rule. (See fig. 57.)

(d) Allowing for thickness of the remaining stock S (fig. 57①), clamp straightedges or ripping fence to table at angle determined by parallel rule.

(e) Make first cut with saw set for a $\frac{1}{16}$-inch cut and take consecutive cuts, raising saw about $\frac{1}{16}$-inch at a time, until the desired depth is reached. Make sure the depth of each cut does not exceed one-half the length of saw teeth. Move stock slowly and steadily over the saw. The stock will be cut to an elliptic curve. (See fig. 58.)

Figure 58. Cutting contour by oblique-sawing method.

(f) Finish contour with sharp gouges or a core-box plane and with scrapers and sandpaper.

(2) *Sighting.* A contour can be cut more completely by sighting than with the parallel rule, but greater caution is required to avoid cutting beyond contour line. Proceed as follows:

(a) Draw contour on end of stock and place stock behind saw. Sighting from front of saw and approximately on a level with the table, move stock until silhouette of saw just covers one side of contour.

(b) Place straightedges along side of stock and clamp them in position.

(c) Make the first cut as in parallel-rule method. With each successive cut reverse the stock end for end so that both sides are cut uniformly. Remove the remaining center portion last and finish as in parallel-rule method. Complete and accurate removal of stock is possible only if saw blade and contour diameters are identical. Then stock is passed over the saw in a direction perpendicular to its rotation and the final cut equals the radius of the blade. Figures 59 and 60 show how to cut common moldings, using the sighting method.

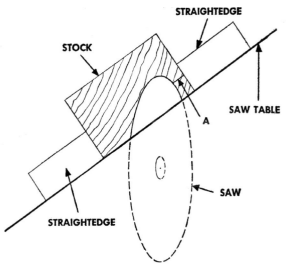

Figure 59. Cutting scotia molding. Tilt table to make undercut contour. Make straight cut at A last.

20. Cutting Wedges and Tapers

a. CUTTING WEDGES. Any number of similarly shaped wedges can be cut by using a template:

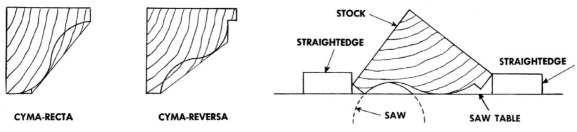

Figure 60. Cutting cyma-recta and cyma-reversa moldings. Miter face to remove waste material, then cut curved contours. Finish molding as in kerf method.

(1) Cut the shape of the wedge into edge of template stock. (See fig. 61.)

(2) Move ripping fence so the template just passes between saw and fence. Use a ripsaw or combination saw.

(3) Fit stock into notch in template and move stock and template past the saw.

(4) Reverse stock end for end for each new wedge.

b. TAPERING SQUARE LEGS. Square legs can be tapered by using the template shown in figure 62①. Proceed as follows:

(1) Set the ripping fence at a distance from the saw equal to combined width of the long template member and the leg stock, so the saw just clears the leg when stock is placed against the template. (See fig. 62②.)

(2) Cut the taper on two adjacent sides with the end of the stock in the first notch. (See fig. 62③.)

(3) Cut tapers on other two sides with end of stock in second notch. (See fig. 62④).

(4) Use a smoothing plane and sandpaper to eliminate the slight difference in length between tapers cut in the two notches.

21. Tenons

The best method of cutting tenons is with a single- or double-end tenoner. However, they can be made with a circular saw, using a combination saw, a dado head, or both ripsaw and crosscut saw. When cutting tenons with a saw, care is necessary to insure that shoulders meet at the correct angle and that tenons fit snugly into mortises. (See par. 50.) Cut tenon shoulders first, then make cheek cuts.

a. SQUARE-SHOULDER CUTS. To cut shoulders on square-shouldered tenons—

(1) Cut stock to finished width, thickness

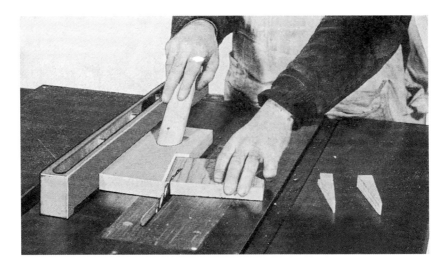

Figure 61. Using template to cut wedges.

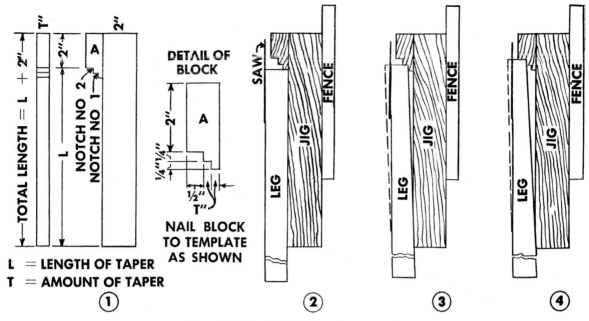

L = LENGTH OF TAPER
T = AMOUNT OF TAPER

DETAIL OF BLOCK

NAIL BLOCK TO TEMPLATE AS SHOWN

Figure 62. Template for tapering square legs.

and length, using a crosscut or combination saw.

(2) Lay out tenon on end of stock. Usually, tenons are one-half the thickness of the stock.

(3) Raise or lower saw or table for correct depth of shoulder cut and clamp it securely.

(4) Set miter cut-off gauge at right angles to saw.

(5) Use ripping fence, stop block, or clearance block to control length of tenon.

(a) *Ripping fence.* Set ripping fence so the distance from left cutting edge of saw to fence equals the length of the tenon, and clamp it securely. Hold tenon end of stock against the fence and push stock into saw. (See fig. 63.)

(b) *Stop block.* Remove fence and place a universal gauge or a miter cut-off gauge to

Figure 63. Cutting square-shouldered tenon, using fence as stop.

71

left of saw. Fasten stop block to gauge so stock projects to the right of the saw and the distance from left cutting edge of the saw to right end of stock equals length of tenon. If shoulders on two ends of a piece are to be of different lengths, a clearance block can be used in combination with the fence. Set block back far enough so stock will clear saw when fence is used.

(c) Clearance block. Fasten clearance block to the fence as shown in figure 64. Hold stock firmly, as there is a chance of error from moving along the gauge, since stock is free to move in either direction after it passes the block.

(3) Holding stock firmly against one of the gauges, cut first shoulder.

(4) Move stock to the other gauge, keeping the same working edge against the gauge. Cut second shoulder.

(5) Reset stops if necessary and cut tenons on the second end.

Note. Use a band saw to cut edge shoulders at an angle.

c. CHEEK CUTS. Using a ripsaw or combination saw, make tenon cheek cuts as follows:

(1) Remove the universal or miter cut-off gauge and secure the table.

(2) Lay out cheek lines of the tenon care-

Figure 64. Cutting shoulder of tenon, using clearance block on fence as a guide.

(6) Use same set-up to cut shoulders on all identical tenons. Reset saw and fence or stop if length or thickness of stock varies.

b. ANGLED SHOULDER CUTS. Tenons with shoulders cut at an angle are often used. To make angled shoulder cuts—

(1) Fasten a universal gauge on each side of the saw at correct angle to give desired shoulder cut. Make certain both gauges are at exactly the same angle to insure that shoulder kerfs coincide with each other on opposite sides of each tenon.

(2) Attach a stop on each gauge, setting both stops at the same distance from the cutting edge of the saw.

fully with a marking gauge, then adjust the fence so the saw will cut just the cheek line.

(3) Make cheek cuts, guiding stock with a push stick about 5 inches wide. (See fig. 65.) If tenon has square shoulders, cut end of push stick at right angles to edge; if face shoulders are at an angle, cut end of push stick at the angle which will brace stock in the correct position. Although it is not essential, a feather board fastened to table, as in figure 66, helps hold stock securely against the fence and also protects the operator. Use a jig band-sawed to correct contours to hold curved stock in position while cutting tenons.

(4) Generally, all work is done from the

Figure 65. Cutting tenon cheeks.

Figure 66. Using feather board to steady stock when cutting tenon cheek.

marked surface or edge; however, either face or edge can be held against the ripping fence if the stock is dimensioned accurately. If there is any doubt as to uniformity in thickness of the stock, cut all work with the same working face against the fence, readjusting fence for the second cheek cut.

Note. Both cheek cuts can be made in one operation by using two saws at once. Mount both

saws on arbor with a filler collar to hold them the correct distance apart. In determining size of the collar, allow for set of the saw teeth; otherwise, the tenon will be too thin. Use paper or cardboard fillers for fine adjustments of space between saws.

d. USING DADO HEAD. Cheek and shoulder of a tenon can be cut in the same operation with a dado head. However, thickness and width of stock must be uniform. Proceed as follows:

(1) Assemble dado head with only one outside cutter and place it on the side of the saw toward the stock. Use sharp cutters in order to leave the bottom of the cut smooth. If possible, use enough cutters so width of dado head equals length of tenon.

(2) Attach a wood face board to fence for protection of saw. Set ripping fence so distance from the fence to left cutting edge of outside cutter equals length of the tenon.

(3) Adjust dado vertically so depth of cut equals amount of stock to be removed from tenon cheek.

(4) Slide stock across saw, cutting one side of the tenon. Turn stock over and cut other side.

Note. If dado head is not wide enough to cut one side of a tenon with one pass, make as many passes as required, resetting fence each time.

22. Special Jigs and Fixtures

Many pieces too awkward to handle in the regular way can be worked with a special jig or fixture mounted on the table. For example, for chair rockers, which require trimming on two ends at different angles, locating pins can be used to place the rocker for the first cut. Then reverse the rocker and insert pins for second cut. Whenever possible, arrange jigs and fixtures so the miter or any other angular crosscut is made from the short side toward the long side. This leaves all the fuzz on the waste block.

23. General

The overarm saw (fig. 67) can be used for ripping, crosscutting, mitering, or beveling by changing the position of the arm, yoke, or motor. Dado heads, shapers, or routers can be mounted on the machine, permitting a wide range of operation with these special cutters.

24. Crosscutting

For straight crosscutting, place the arm at right angles to the guide strip by setting the miter latch in the column slot at zero (0°) position. Lock the arm with the arm-clamp handle. Place material on work table against guide strip and make the cut (fig. 68), then return the saw blade behind the guide strip.

25. Mitering

For miter cutting (fig. 69), pull the saw carriage as far back as it will go and release arm clamp and miter latch. Swing the arm to the desired angle on the miter scale; for 45° miter cuts, right or left, or straight cut-offs, locate arm latch in the proper column slot. For any other position, lock the arm with the clamp handle and proceed with cutting.

26. Ripping

For ripping, the arm must be clamped in crosscut position. Pull the entire motor carriage to the front of the arm, lift the rip latch and release swivel-clamp handle, and revolve the motor yoke under its carriage. Holes drilled in the yoke receive the rip latch and hold the motor in cut-off, in-rip, or out-rip position. Using rip scale on the arm, locate position desired, lock swivel clamp, and secure rip lock on the arm. Adjust safety guard so infeed end almost touches the material. Lower kick-back device on opposite end so the sharpened points rest on the material. Keep material against the guide strip and feed it evenly into the saw blade. (See fig. 70.) Do not feed material into kick-back end of the saw guard.

27. Angle Cut-Off

For an angle cut-off, tilt the motor with machine in crosscut position. Raise column with the crank, release bevel latch and bevel-clamp handle, and turn motor in the yoke. Use the dial scale to set desired angle. The bevel latch locates 45° or 90° positions; if any other angle is desired, use the bevel clamp to hold motor rigidly in position. Lower column enough to make sure the saw blade cuts through the material and proceed as for crosscutting. (See fig. 71.)

28. Compound Miter

Compound miter cuts are combined bevel and miter cuts. To cut a compound miter, set up machine for bevel cut-off. Release arm latch and clamp handle, and swing the arm into desired miter position. (See par. 25.) Follow normal crosscutting routine. Bring saw blade across material steadily, giving it a chance to cut. (See fig. 72.)

29. Bevel Ripping

For bevel ripping, tilt the motor with the machine in rip position. Raise column by rotating the handle. Release bevel latch and bevel-clamp handle and turn motor in the yoke. Locate 45° position with the bevel latch; lock the motor at any other angle with bevel-clamp handle. Adjust guard on infeed end so it almost touches the material, but do not adjust kick-back device. Use a push stick to prevent kick-back of material. Feed stock steadily into saw. (See fig. 73.)

30. Dadoing

For dadoing, remove saw blade from motor arbor and mount dado head. To determine depth of cut, lower column until dado touches top of material, then remove material and continue to lower dado head by rotating crank handle. Each complete turn of handle lowers column $\frac{1}{16}$-inch. Cut straight dadoes using crosscutting procedure and angle dadoes using mitering procedure. (See fig. 74.) Make wide cuts by moving dado back and forth over material. The dado head can be operated in either direction.

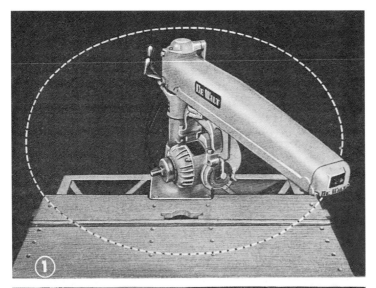

① *Circular swing of arm.*

② *Circular swivel of yoke.*

③ *Circular rotation of motor and arbor.*

Figure 67. Overarm saw.

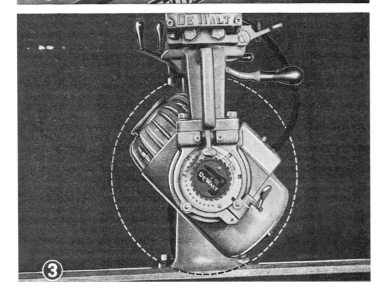

Figure 68. Crosscutting with overarm saw.

Figure 69. Miter cutting.

Figure 70. Ripping.

Figure 71. Angle cut-off.

Figure 72. Compound miter cutting.

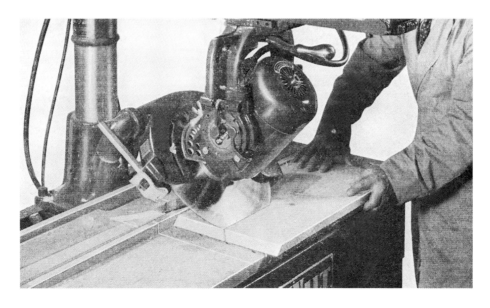

Figure 73. Bevel ripping.

Figure 74. Dadoing.

31. Plowing

To plow, replace saw blades with dado head. Place dado in desired position, clamp rip lock on arm, lower dado to depth wanted. Adjust safety guard so that infeed end almost touches material, then set kick-back device on opposite end. Feed material steadily, holding it against guide strip for best cutting results. (See fig. 75.)

32. Rabbeting

Make rabbet cuts with dado head, using the set-up described for plowing. Set motor in place behind guide strip, release bevel latch and

Figure 75. Plowing.

clamp, and raise the column until motor is in vertical position. Clamp dado securely in 90° slot and lower dado head to desired depth. Bring out dado in front of guide strip for width wanted. Adjust dust spout on safety guard. Feed material evenly and steadily, keeping it against guide strip for accurate work. (See fig. 76.)

33. Cutting Tenons

To cut tenons, insert spacing collars between dado blades and set motor as for rabbeting. Use a wider guide strip so that guide fence is about 3 inches above the table top. Pull tenon cutters forward in front of guide fence to get desired cut. Use a wooden push block to feed material past cutters. (See fig. 77.) Make sure table top is level.

34. Shaping

The overarm saw can be used for shaping when standard shaping equipment is not available, but it is not fast enough for extensive work. To use the saw for shaping, mount the shaper cutter head on the saw arbor. Have motor in same position as for rabbeting. Vertical and horizontal adjustments on machine permit any part of the shaper cutter to be used or eliminated. Adjust machine so desired form or shape is profiled on the end of the material. Lock motor carriage by securing rip lock on arm. For best results, feed material into shaper cutter along the grain. (See fig. 78.)

35. Routing

To use the overarm saw for routing, remove front saw arbor and screw a router bit into the back end of the motor shaft. Use a bit with a ½-inch right hand No. 14 threaded shank. Fix machine in same position as for shaping. Move guide strip toward rear of work table for more working space. Lower router bit into material by revolving the column elevating handle. Lock router bit in one position by securing rip lock on arm. Do freehand routing by moving material or bit. When using the latter method, nail material to work table, release rip lock, and move router bit by swinging arm and moving motor carriage as necessary. Templates also can be used. (See fig. 79.)

Figure 76. Rabbeting.

Figure 77. Cutting tenon.

Figure 78. Shaping.

Figure 79. Using router bit and template for
routing work.

36. General

Wood jointers are used primarily for surfacing and edging but are used also for rabbeting, chamfering or beveling, and tapering. The jointer and the surfacer or planer are similar in cutting action. One major difference, the absence of pressure bars and rollers on the jointer, makes it possible to process stock on this machine so that any warpage is eliminated. However, this is done at the expense of board thickness or width, so only minor warpage can be eliminated without excessive waste of lumber. After one face has been reduced to a plane surface, the lumber is run through a surfacer to bring the other face into a parallel plane. Similarly, after one edge of a board has been jointed, the board may be ripped into strips of uniform width when the jointed edge is kept against the saw fence. The planer cannot take warped lumber and process it into an acceptable product unless one face has been jointed. The work is done by a cutting head, frequently having three knives, which revolves at 3,600 to 4,500 rpm. Jointers vary in size from models with a 4-inch cutter head to models with a 36-inch head. Figure 80 shows a typical jointer with base, front and rear tables, cutter head,

power attachment, and fence. The work is fed over the cutter head by hand, and its quality depends to a large extent on the skill of the operator.

37. Surfacing

Boards are surfaced on a jointer to make one face a plane surface.

a. PROCEDURE. To surface a board—

(1) Adjust front table to make a 1/32-inch cut.

(2) Set fence at right angles to the table and close to the right edge, to provide maximum working surface.

(3) Start machine. When cutter head is revolving at maximum speed, hold stock firmly against front table and fence and push it over the cutter head. (See fig. 81.)

(4) As soon as the front part of the stock passes the cutter head, hold it down on the rear table. Since the rear table determines the plane of cut, stock must be held against it firmly to insure a true plane. Joint cupped stock with concave surface down.

b. LIMITATIONS. Width of the cutter head limits the width of board that can be surfaced.

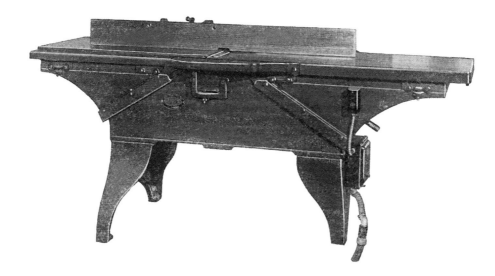

Figure 80. The jointer.

Figure 81. Surfacing board.

38. Edging

Edging straightens one edge of a piece of lumber and facilitates later ripping, crosscutting, or gluing.

    a. UNWARPED STOCK. If the edge is fairly

straight, hold the board firmly against fence and front table and push it over the cutter head to the rear table. (See fig. 82.) Repeat until board edge is perfectly flat.

    b. WARPED STOCK. (1) Joint warped stock

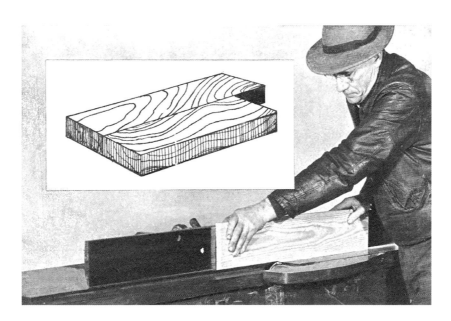

Figure 82. Edging unwarped stock.

with concave surface down. Pass board over cutter four or five times to reduce the rear tip. (See fig. 83.) Reverse board and cut the other tip, then pass entire board over cutter in the usual way. This permits knives to start cutting gradually instead of jabbing into the edge and possibly splitting the board.

(2) If the final shape of the board makes it more practical to joint the convex edge, hold board in the approximate plane of the finished edge. Pass board over cutter head as often as necessary to remove irregularities.

(3) If a board is badly warped, cut edge approximately straight on a band saw and finish on a jointer.

39. Rabbeting

To cut rabbets on a jointer, set fence so its distance from cutter edge equals width of rabbet and adjust front table so cut equals depth of rabbet. Then proceed as for ordinary surfacing. (See par. 37a.)

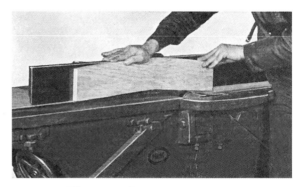

Figure 83. Jointing crooked stock.

40. Chamfering and Beveling

To chamfer or bevel stock on a jointer, set fence at the required angle and hold stock firmly against it (figs. 85 and 86) ; any gap between stock and fence results in a change in angle of the bevel. The technique of these operations is similar to that of surfacing or edging.

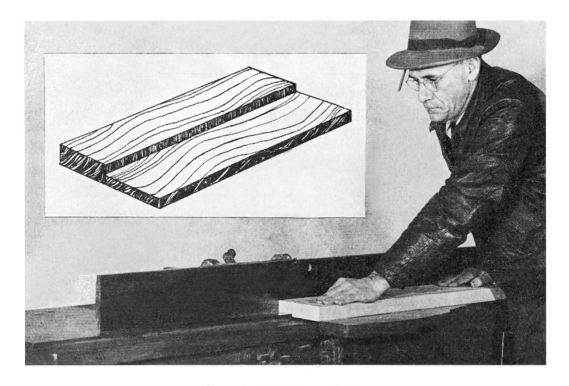

Figure 84. Rabbeting on jointer.

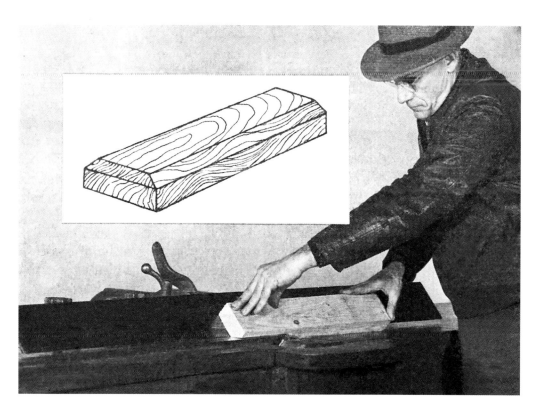

Figure 85. Chamfering on jointer.

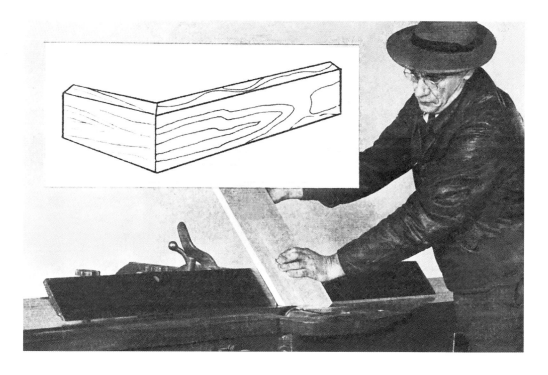

Figure 86. Beveling for miter joint on jointer.

41. General

The single surfacer or planer (figs. 87 and 88) is used to remove saw marks and make stock uniformly thick throughout its length. Best results are obtained if one side of stock is surfaced on a jointer before the other side is planed. The single surfacer planes only one side at a time. It can handle stock up to 7 or 8 inches thick. Double surfacers, which plane both sides of a board simultaneously, are generally used only in large planing mills. Single surfaces are made in sizes ranging from a 12-inch to a 48-inch bed width.

42. Principal Parts

Principal parts of the surfacer are—

a. SUPPORTING FRAME. The supporting frame is a heavy casting which supports the other parts of the machine. Its weight helps decrease vibration. It must be bolted to a solid level foundation.

b. BED. The bed of the surfacer is a flat, table-like surface on which the stock is laid and automatically carried forward for the planing process. The conventional bed has three sections, front, center, and rear, with rollers on either side of the center section to move the

Figure 87. Front view of 18-inch single surfacer.

Figure 88. Left side of 18-inch single surfacer showing feed works and grinding equipment attached.

stock forward. The lower infeed roller, between front and center sections of the bed, runs the full width of the table and is set a little higher than the bed. The outfeed roller is about 10 inches to the rear, between the center and rear sections. It is set at the same height as the infeed roller. Bed and rollers can be moved vertically as a unit by a raising and lowering device.

c. RAISING AND LOWERING DEVICE. The raising and lowering device is usually operated manually by a screw having a wheel and handle projecting in front of the frame. One revolution of the wheel ordinarily changes bed elevation $\frac{1}{16}$ inch. A scale and pointer mounted in the front show the thickness to which the surfacer is set. In some surfacers, the bed is raised or lowered by a handle which moves a wedge-shaped casting forward or backward. The upper edge of the wedge has a machined surface on which rests a similar reversed-wedge casting attached to the bed. Since the bed can only move vertically, it is raised when the lower wedge is moved forward and is lowered when the wedge is moved back.

d. FEED ROLLS. Another set of infeed and outfeed rollers, located directly over their respective counterparts in the bed, is carried in the upper section of the surfacer. (See fig. 89.) The upper infeed roller is adjusted to about the same height as the chip breaker and knives, while the upper outfeed roller is lower. The outfeed roller is usually a floating unit, relying on its weight for enough pressure on the board to continue the feed.

e. CUTTER HEAD. The cutter head is in the center of the top section. Usually it is a round cylinder, slotted to hold the cutting knives, which are held in the head by setscrews. The number of knives varies with different machines; three knives, extending the width of the machine, are frequently used. The best knives are high-speed steel about ⅛-inch thick and 1¼ inches wide.

f. CHIP BREAKER. The chip breaker, located directly in front of the cutter head, prevents excessive chipping by pressing lumber down while it is planed and forcing chips to break into small pieces. One type is solid, hinged so it can rise or drop when irregular boards are planed. Another type has many sections which move up and down individually. This type is

effective when boards of different thicknesses are being processed. Surfaces with sectional chip breakers also have a sectional upper infeed roller made up of 1-inch wide rollers.

g. PRESSURE BAR. The pressure bar, directly behind the cutter head, extends across the width of the machine. It has a smooth lower surface which holds the lumber firmly in place after cutting, preventing chattering and insuring smooth work.

h. DRIVING MECHANISM. The feed works are usually powered by a motor attached to one end of the cutter head. In some older installations, however, power comes from a central source, and in this case a belt is connected to a pulley on the end of the cutter-head shaft. Power is transmitted to other moving parts by gears, belts, or chains, sized to transform the speed to the desired rate. Most surfacers have controls for varying the rate of feed to suit the type of material processed.

i. GRINDING ATTACHMENT. A grinding attachment can be mounted on the surfacer to sharpen cutting knives automatically while they are assembled in the head.

43. General Operating Instructions

Examine stock before processing it to make sure it is free from foreign material, such as nails and shot, which might chip the cutters. Before surfacing, run boards over the jointer to produce one straight side. (See par. 36.) When surfacing a piece such as a square table leg, joint and square two adjoining sides before planing. Make sure that a jointed surface rests on the machine bed.

a. ADJUSTMENT OF SCALE. When machine knives are ground in place, adjust the scale setting after each grinding. Otherwise, a scale setting of 1 inch, for example, will produce stock surfaced 1 inch plus the amount ground off the knives since the last scale adjustment. Run a piece of scrap lumber through the machine to determine whether the scale is accurate.

b. RATE OF FEED. The cutter head makes about 3,600 rpms. The average work is fed at about 25 feet a minute, with the cutter head operating at maximum speed. If the motor slows down, reduce either rate of feed or amount of cut.

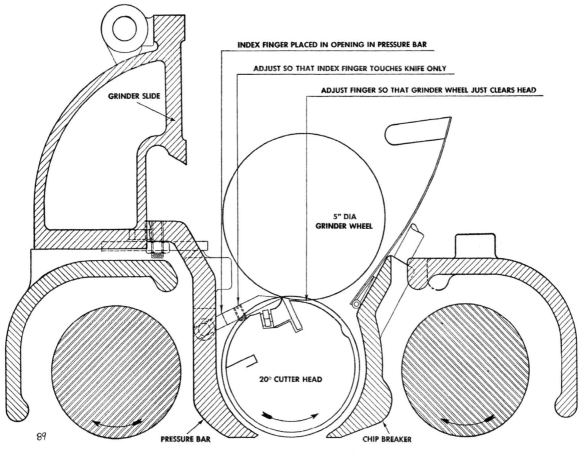

INDEX FINGER PLACED IN OPENING IN PRESSURE BAR

ADJUST SO THAT INDEX FINGER TOUCHES KNIFE ONLY

ADJUST FINGER SO THAT GRINDER WHEEL JUST CLEARS HEAD

GRINDER SLIDE

5" DIA
GRINDER WHEEL

20° CUTTER HEAD

89

PRESSURE BAR

CHIP BREAKER

Figure 89. Cutter head, pressure bar, chip breaker, and
upper feed rollers of a single surfacer.

c. AMOUNT OF CUT. The maximum cut for average work is $\frac{1}{16}$ inch. A wide piece naturally requires more power for cutting than a narrow piece, and hardwoods offer more resistance than softwoods. If motor cannot make a $\frac{1}{16}$-inch cut at maximum speed, reduce depth of cut. If several cuts must be made to reduce lumber to the desired thickness, turn the board over for each cut, removing the same amount from each side. Since moisture content at the center of a board is usually different from that at the surface, the board has a tendency to cup unless balance is maintained.

d. DIRECTION OF CUT. When practicable, run stock through surfacer at a slight angle to reduce danger of chipping.

e. LENGTH OF BOARDS. Do not try to surface stock less than 2 inches longer than the distance between front and rear rollers, because the outfeed roller will not pick up a shorter piece.

Feeding short pieces one behind the other to push them through the machine is not good practice, because boards have a tendency to jump and the finished work is not smooth.

f. FEEDING. Use entire bed of the planer. Using one side more than the other shortens the life of the machine. Feed stock so the knives strike approximately with the grain, which rarely is parallel to the surface to be planed. If direction of grain cannot be determined visually, place a few drops of stain on the edge of the wood; the stain will follow the grain. Do not confuse grain with pattern or texture.

g. USING CARRIER. Use a carrier to taper pieces with the surfacer and to process thin stock.

(1) *Tapering.* Prepare carrier for tapering (fig. 90) by cutting a board slightly longer than the stock to be processed and tapering it to the

desired angle. Fasten a stop block to the thick end. Lay the work against the stop and run stock and carrier through surfacer, thin end first. Process stock on two adjacent sides. Finish the work by trimming the two ends to form right angles with a line drawn through the center of the piece. If the piece is trimmed to exact length before tapering is begun, plane the first two sides on a carrier tapered to half the required angle. Plane remaining two sides on a carrier tapered to the full angle.

(2) *Processing thin stock.* Use a carrier if

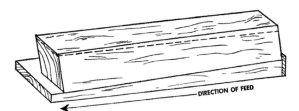

Figure 90. Surfacer carrier for tapered legs.

stock is to be planed down to less than 1/4-inch thickness. Build carrier by attaching a thin stop to one end of a 1-inch board which is slightly longer than stock to be processed. (See fig. 91.) Set the thin piece on top of the carrier, against the stop, and proceed in the same way as for a single board.

Figure 91. Surfacer carrier for thin boards.

h. SAFETY MEASURES. Be careful in feeding boards of unequal thickness on a machine with a solid upper infeed roller. The rollers may advance thinner boards to the cutter head, but they do not exert enough pressure to keep the high-speed knives from kicking the board back against the operator.

44. General

The wood shaper should be operated only by thoroughly experienced mechanics. Serious injuries may result if all precautions are not taken. Always follow standard production practices; do not experiment with this machine. If possible, use the protecting guards; if this is impracticable, construct the jig so as to protect the operator. The shaper (fig. 92) is designed primarily for cutting moldings but can also be used for rabbeting, grooving, beading, and fluting. The shaping tool may be either a solid-head or flat-knife cutter.

a. SOLID-HEAD CUTTER. The solid-head cutter is a solid disk of steel with the shape milled on the circumference. Sections are cut away to leave multiple wings or blades which cut the desired pattern. (See fig. 93.) The center of the cutter is drilled to fit over the shaper spindle.

b. FLAT-KNIFE CUTTER. The flat-knife cutter is made from two flat pieces of shaper steel. (See fig. 93.) Top and bottom edges are tapered to fit in grooved collars (fig. 93) which hold the knives on the shaper spindle. (See fig. 93.) Blades can be ground to any desired shape in the shop. Both blades must be identical in shape and weight and must project the same distance between collars. The nut holding knives between the collars must be tight. Failure to observe these precautions will result in defective work and damage to bearings. The vertical shaper spindle revolves at high speed (frequently 10,000 rpm), and the centrifugal force developed emphasizes any irregularity.

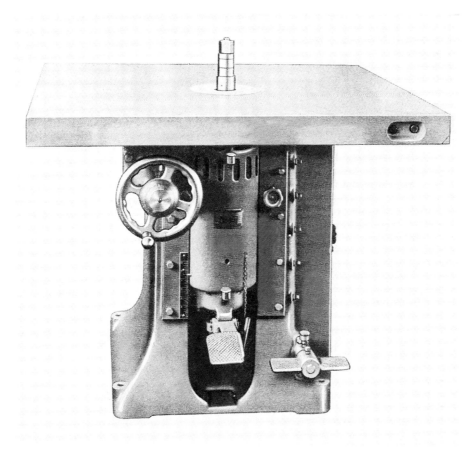

Figure 92. Wood shaper—front view.

THREE-WING SOLID CUTTER

FLAT KNIFE

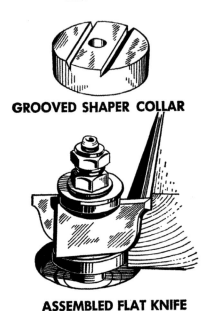

GROOVED SHAPER COLLAR

**ASSEMBLED FLAT KNIFE
SHAPER HEAD**

Figure 93. Shaper cutters.

45. Operations

a. SHAPING STOCK WITH STRAIGHT EDGES
AND EXTERNAL ANGLES. The shaping operation
is simplified by fixing a straightedge to the
shaper table for use as a fence. Saw out a circular portion of the fence so it fits around the
cutters, and attach fence to table.

(1) To shape part of an edge (fig. 94),
hold stock firmly against the fence and force it
past cutting head.

(2) Feed stock in the direction opposite
to the revolving cutters and at a speed which
does not slow the spindle. Slower feed usually
means smoother work. Adjust spindle for

proper knife height, using the handwheel on
the side of the machine.

(3) Whenever possible, make cut so long-
est projection of knives is toward bottom of
stock.

Caution: If entire edge is to be shaped, the
fence will not support stock after it passes cutter. Therefore, before shaping entire edge, tack
a strip of wood about ⅛-inch thick to the portion of fence beyond cutters. Adjust fence so
cutters waste an amount of stock equal to the
thickness of the offset strip.

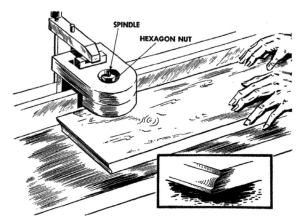

Figure 94. Shaping stock, using fence as guide.

b. SHAPING STOCK WITH CURVED EDGES AND
INTERNAL ANGLES. Shape stock with curved
edges or internal angles, using the revolving
collars as guides. Regulate depth of cut by adjusting the distance which knives project beyond the collars, or by using collars of different
diameters.

(1) Clamp a guide strip on the shaper
table about ¼-inch from the cutters. This af-
fords a firm fence for starting the feed. After
cutting starts, swing stock away from the strip
and use the collar to guide the work. (See fig.
95.)

(2) To shape only part of the edge, draw
stock around the cutter head with the uncut
part against the collar.

(3) To shape entire edge, make a metal
or wooden template with the same contours as
edge to be shaped. Attach template to top of
stock with tacks or other fastening and push
stock past cutters, holding edge of the template
against the collar to guide the cut.

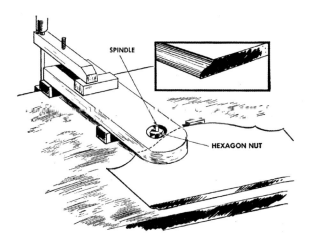

Figure 95. Using collar and guide strip to shape stock with internal angles.

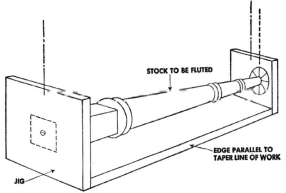

Figure 96. Jig for fluting on shaper.

c. FLUTING AND BEADING. Use a jig for fluting or beading on a shaper. (See fig. 96.) Make sure the edge of the jig running against the fence or collar has the same contour as the piece to be fluted. Mark off spaces for each cut and fasten stock securely in jig.

46. Safety Measures

Be sure that stock is flat and rests firmly on the table. A firm, steady feed is necessary for good work. Always attach a wood spring to hold stock near the cutters. (See fig. 95.) Allow only experienced personnel to shape curved articles such as certain types of chair arms, where the feed moves both horizontally and vertically.

47. General

The chief use of the band saw (figs. 97 and 98) is cutting curved edges; however, it is used also for resawing boards and for bevel ripping.

a. BLADE. The band-saw blade must be sharp and accurately set to cut in a straight line. The width of the blade used depends on the radius of the curves to be cut. If the radius is short, use a narrow blade and make saw kerfs in the waste to lessen strain on the saw. If the radius is long, use a wider and stronger blade.

b. TABLE. The band-saw table can be tilted to an angle of 45°, but for conventional sawing it is usually horizontal, with the saw blade perpendicular to it.

48. Operation

a. GENERAL. (1) The bottom surface of stock must be flat. If the piece wobbles, work will be inaccurate and the saw blade may kink.

(2) Clamp the top guide of the machine about ¼-inch above stock. If necessary, adjust

Figure 97. Band saw with guards swung back.

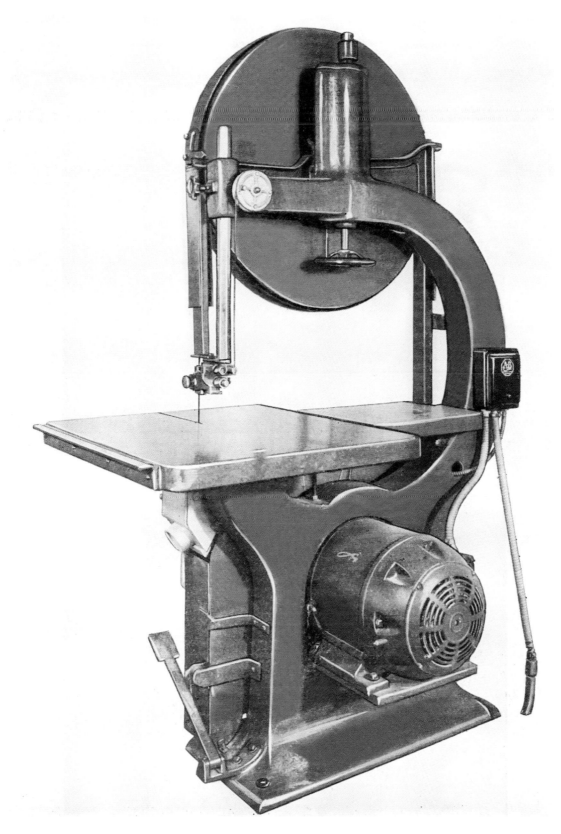

Figure 98. Rear view of band saw.

the angle of the top saw wheel to make sure the saw blade rides against the guide wheels without pressure. Always operate machine by hand while making adjustments.

(3) Guide stock along the line marked on the face of the board, holding it loosely and taking care not to crowd the saw. (See fig. 99.)

(4) If cutting has to be stopped, let the saw cut its way out through waste instead of backing it out. However, if backing out is necessary, be sure to follow the exact cut; otherwise the saw may pull off the wheels.

Figure 99. Cutting straight kerfs on band saw to prevent binding in subsequent cutting of small-radius curves.

Figure 100. Resawing on band saw, using fence.

b. RESAWING. Resaw boards to reduce their thickness. (See fig. 100.) If a ripping gauge is not available, clamp a wood fence to the table for use as a guide. If possible, surface the bottom edge and both sides of stock before resawing. If resawing before sides are surfaced is necessary, mark the top edge of the board to show were the cut should be made. Be sure to joint the bottom edge. Then hold a right-angle block against the side of stock with one hand and regulate the feed with the other. (See fig. 101.) This permits changing feed angle if the board is warped.

c. BEVEL RIPPING. Use the band saw for bevel ripping and for cutting angles on irregularly shaped stock by tilting table to the desired angle and processing stock as in vertical sawing. When bevel ripping, set the fence on the lower side of the saw; when cutting irregular shapes, use either side of the table. (See fig. 102.) If upper side of the table is used, the larger surface remains on the table, lessening the possibility of tipping.

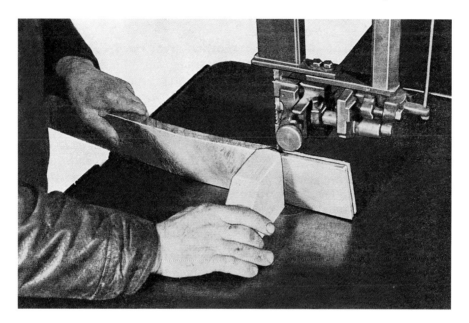

Figure 101. Resawing warped board on band saw, using square block.

Figure 102. Cutting angles on band saw.

49. General

The hollow-chisel mortiser (fig. 103) is designed to cut a recess in wood to receive a matching tenon. The cutting tool consists of a bit which revolves inside a square hollow chisel. The tool is forced into the wood by a treadle or an automatic mechanism. The bit is set to project about 1/8 inch below the chisel. It cuts away most of the wood, disposing of shavings through an opening on one side of the chisel. The chisel does not revolve, but cuts the sharp corners required to complete a rectangular mortise. The mortiser can stand hard continuous use if the chisel edge is protected and the bit correctly adjusted. The machine must be operated with care and the operator must keep his hands away from the chisel and revolving bit.

50. Operation

a. GENERAL. When a single piece is being processed, or when mortise size or location varies between pieces, each mortise must be marked out. When a number of similar pieces with the same mortise size and spacing are being processed, one sample lay-out is enough, since the use of stops, clamps, table adjustments, and guides will make possible identical processing of all pieces. A fine pencil line is satisfactory for most work. For exceptionally precise work the mortise outline can be made with a knife cut.

b. MACHINE ADJUSTMENTS, STOPS, AND CLAMPS. (1) The table has the following adjustments:

(a) Vertical travel (18 to 24 in.). Table height setting is governed by dimensions of the stock and the stroke setting of the sliding head. A proper combination of these factors results in the desired mortise depth.

(b) Cross travel (2 to 4 in.). Setting is governed by mortise location with respect to the stock edge. This adjustment in combination with the longitudinal travel of the table is useful in moving stock for the successive cuts necessary in wide or long mortises.

(c) Longitudinal travel (10 to 14 in.). (See (b) above.)

(2) The sliding head has a maximum vertical travel of approximately 4 inches. This travel may be shortened or regulated by a change in the linkage from the foot treadle, or by use of the head stop nuts.

(3) The fence is adjustable for height and serves to hold or guide the stock edge. In machines with stationary tables, the fence is movable (cross travel) and is used for the same function as the table cross-travel adjustment previously described.

(4) The stock clamp operates by screw action from the front of the table and has a maximum capacity of 6 inches. This is used to clamp stock against the fence during the mortising action.

(5) The spacing gauge is adjustable vertically to allow its use with varying table heights. Stops along the length of the gauge are adjustable for stock length, and serve to regulate the length and spacing of mortises.

c. CUTTING THE MORTISE. (1) Limitations of the machine. Certain limitations of the machine influence the sequence of mortising operations as well as mortising technique. Whenever possible mortising should be done while the stock is still in rectangular form. If a mortise must be cut in shaped or turned stock, use a carrier or jig to hold the stock. Machine size may limit size of cut. In this case the desired mortise depth or width can be obtained by repeated partial cuts. The nature of the cutting tool imposes limitations on its use. Since most of the cutting is done by the revolving bit within the chisel, the bit center must be engaged solidly in the wood to prevent deflection of the cutting tool. Proper bit size and proper spacing of cuts by table or stock movements result in correct overlapping of cuts. These limitations of the cutting tool make it necessary for at least three-fourths of the bit capacity to be used for end cuts.

(2) Procedure. The procedure in cutting a mortise, when the mortise width is a standard bit size, is as follows:

(a) Adjust table height for desired mortise depth.

(b) Center the table for longitudinal travel.

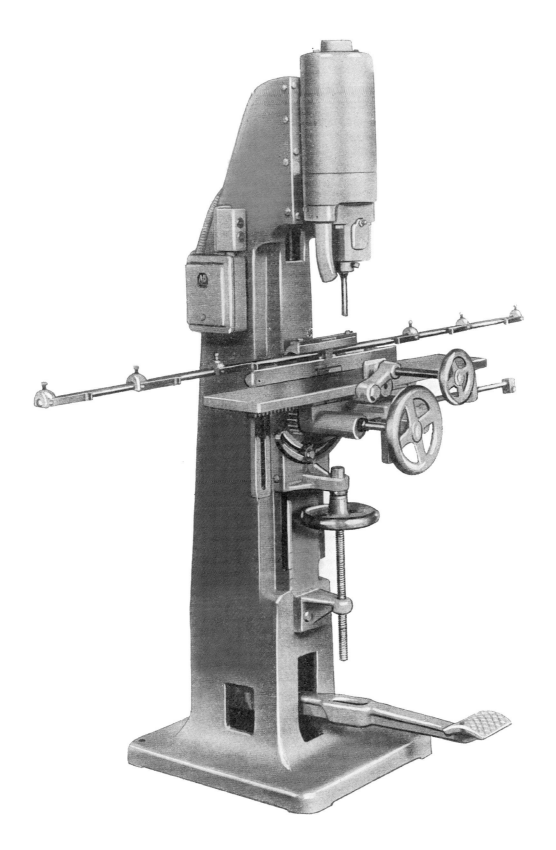

Figure 103. Hollow-chisel mortiser.

(c) Adjust table cross travel so the chisel cuts along the marked mortise area when stock is clamped against the fence.

(d) Make first cut at end of the mortise.

(e) Move stock and make second cut so that it just overlaps the first. Continue moving stock and making cuts until the marked-out area has been removed. Stock is moved for successive cuts by using the longitudinal travel adjustment of the table, or by moving stock along the fence.

> **Caution:** If length of mortise is such that the last cut will utilize less than three-fourths of the bit capacity, make next to last cut at mortise end and use the last cut to clean out stock between the cuts.

(f) If it is not possible to cut the desired mortise depth in one operation, additional strokes or cuts must be made until the desired depth is obtained. The depth of cut is determined by machine model and characteristics of wood being worked.

(3) *Procedure for wide mortises.* When the mortise is wider than an available cutting tool, make two cutting sequences along the length of the mortise to obtain the desired width. The limitations of the cutting tool should be kept in mind and chisel size selected to assure that proper overlapping of cuts is possible. Tool sizes vary from $\frac{1}{4}$- to 1-inch by $\frac{1}{16}$-inch sizes. Maximum allowable tool size depends on model of machine. In cutting a mortise $1\frac{1}{8}$ inches wide, use a $\frac{5}{8}$-inch chisel. Two series of cuts, one along each edge of the mortise, give the desired result without any danger of tool deflection.

51. General

A wood lathe (fig. 104) is used for shaping wood. It consists of the bed, supporting stand, headstock, tool rest, and tailstock. Wood is held between headstock and tailstock and rotated against special cutting tools. (See fig. 105.)

a. HEADSTOCK. The headstock is mounted on the left end of the lathe bed. It consists of a short shaft, the bearings and housing necessary to support the shaft, the motor or cone pulley for power transmission to the shaft, and the spur or faceplate attachment to the shaft. All power for the lathe is transmitted through the headstock. The headstock is not adjustable.

b. TOOL REST. The tool rest is a bar attached to the bed of the lathe between the operator and the work. It may be of any size and is movable in all directions. The tool rest provides support

for the operator in guiding tools along the work.

c. TAILSTOCK. The tailstock is a movable assembly on the right end (end opposite the headstock) of the lathe bed. It consists of a short shaft which is adjustable lengthwise a limited distance (4 to 6 in.), and the supporting housing. The housing is movable along the length of the bed. The shaft usually is taper-bored for the cup center. The tailstock supports one end of the work when the spur is used on the headstock.

52. Tools

Special lathe tools include gouges, skew chisels, parting tools, round-nose chisels, square-nose chisels, spear-point chisels, and auxiliary aids such as outside and inside calipers and dividers.

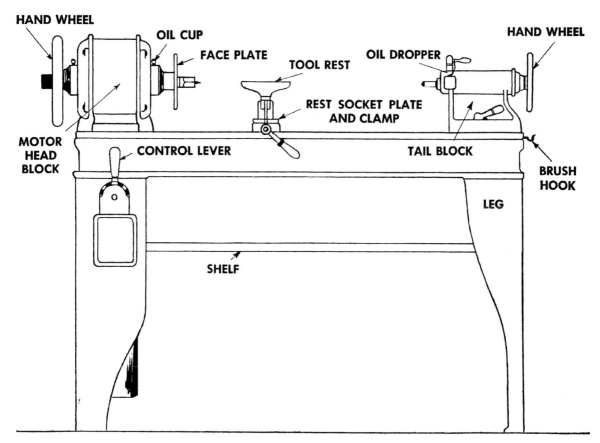

Figure 104. Wood lathe.

a. GOUGES. Use gouges to rough out nearly all shapes. This tool can be used by a skilled workman for almost any operation. Sizes vary from ⅛ inch to 2 or more inches, but ¼-, ¾-, and 1-inch sizes are the most common.

b. SKEW CHISELS. Use skew chisels for smoothing cuts to finish a surface, and for turning beads, trimming ends or shoulders, and making V-cuts.

c. PARTING TOOLS. Use parting tools to cut recesses or grooves with straight sides and a flat or square bottom. These tools are common in ½- to ¾-inch sizes.

d. ROUND-NOSE CHISELS. Use round-nose chisels for rough turning and for forming concave recesses, coves, and circular grooves, mostly in faceplate turning. Common sizes are ⅛-, ¼-, and ½-inch.

e. SQUARE-NOSE CHISELS. Use square-nose chisels for smoothing convex or flat surfaces in faceplate turning. Common sizes are ½- and ¾-inch.

f. SPEAR-POINT CHISELS. Use spear-point chisels to finish insides of recesses and square corners. They are scraping tools and can be made by grinding skew chisels of the desired size.

g. CALIPERS AND DIVIDERS. Use calipers and dividers to measure, mark, and gauge dimensions.

53. Wood-Turning Procedures

Turning operations are performed on stock held between centers or on stock fastened to a faceplate. In the former instance, turning is done primarily along the length of the piece; in the latter instance, turning is done on the face and side.

a. STOCK HELD BETWEEN CENTERS. Turn sides of stock using the following procedure:

(1) Cut stock approximately square and about 1 inch longer and ¼ inch wider than desired finished length. Draw lines diagonally across both ends of stock; the intersection of the lines is the approximate center. (See fig. 106 ①). If end cross section is not a square, estimate the center point, using method shown in figure 106 ② and ③. After end center is located, indent with a center punch or drill a hole ⅛ to ¼ inch deep.

(2) Hold stock centered against spur on headstock of lathe and tap lightly with a wooden

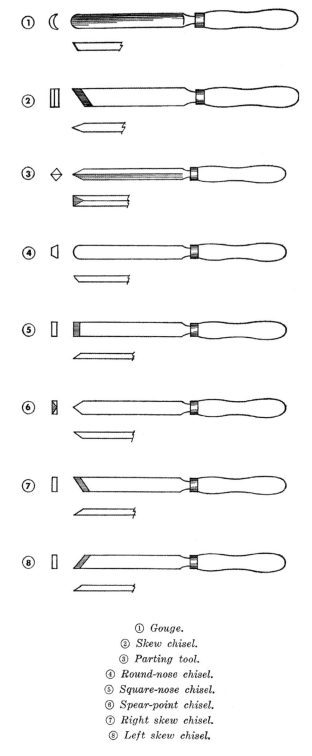

① Gouge.
② Skew chisel.
③ Parting tool.
④ Round-nose chisel.
⑤ Square-nose chisel.
⑥ Spear-point chisel.
⑦ Right skew chisel.
⑧ Left skew chisel.

Figure 105. Turning tools.

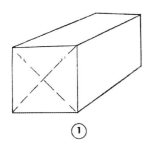 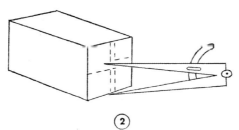 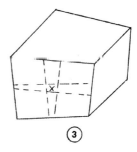

Figure 106. Finding lathe center points.

Note. If stock is not square, set dividers at about half the thickness of stock and scribe lines parallel to edges as in ②. Estimate center inside scribed lines.

mallet to seat it against spur surface. Move tailstock to within ¼ inch of stock end and clamp the tailstock to the bed. Run tailstock spindle out and place center point in center hole on the end of stock. Force center point tightly against the end of the piece, then back off spindle about one-eighth turn of the handwheel and place a drop of oil where the wood touches the metal center. This releases active pressure between centers and allows work to rotate freely without end play. When turning extremely long stock, attach one or more supports to lathe bed to keep stock from bending.

(3) Move tool rest to within ⅛ inch of the work and parallel to it. Set tool rest approximately level with or about ⅛ inch below center line of stock.

(4) Set lathe speed according to diameter and length of the piece. For stock not more than ½ inch square and 12 inches long, use high speed. If length or side dimensions are increased, use second or third speed. Turn stock larger than 4 inches square or more than 3 feet long at low speed.

(5) Use the gouge to reduce stock to a cylinder and for making concave cuts and grooves. Holding gouge handle securely in right hand, place gouge on tool rest with handle lower than tool rest. Steady gouge with left hand by placing the heel of the palm against the front of the tool rest; hold palm on top of the tool for heavy cuts and under the tool for finer cuts. Bring gouge against work, rolling it sideways to make a shearing cut with the side edge of the tool, and move it slowly along stock. (See fig. 107.) Check dimensions frequently with calipers set at least ⅛ inch over finish dimensions until enough skill has been acquired to turn down to finish dimensions at the first attempt.

Figure 107. Roughing out stock with gouge.

105

(6) Move tool rest as work progresses, keeping about ⅛-inch clearance so the hand steadying and guiding tool edge has full support at all times. Adjust angle of rest to conform to contour of piece being processed.

(7) After stock has been reduced to approximate size and shape desired, use the skew chisel for smoothing cuts, V-cuts, convex cuts and for trimming ends and shoulders. Always place chisel on the tool rest before bringing it against the work. Draw chisel back slowly, meanwhile raising the handle slightly until the heel of the bevel on the cutting edge engages the work. Keep face of the bevel tangent to the surface of the work. Hold chisel firmly against tool rest and move it steadily into the cut. (See fig. 108.) Never allow the point to come into the cutting position because it may dig in and ruin the turning. Move chisel laterally in a direction approximately parallel to angle of the tool edge. Reverse chisel to cut in the opposite direction.

(8) In case beads, coves, fillets, or other turning patterns are specified, the parting tool, and other tools listed in paragraph 52, may be used. The cuts made by these tools are mainly of the scraping type.

b. FACEPLATE TURNING. Use following procedure for faceplate turning:

(1) Replace spur on head stock with a faceplate. Fasten work to faceplate by one of the methods below.

(a) Mount stock under 4 inches in diameter and under 2 inches in thickness on a chuck with a screw permanently fastened in its center. Be sure the screw does not interfere with turning.

(b) If faceplate has a wooden surface, fasten work to the wooden face with small headless brads or with glue.

1. If brads are used, countersink their heads so they will not interfere with the turning operation. After turning, remove work from faceplate with a thin chisel. Insert chisel between plate and work, starting at the end grain, and work around the piece until it is entirely loose.

2. If glue is used, fasten the work to the wooden faceplate with a standard glue joint. After turning, cut work from faceplate with a parting tool. Stock can be removed from faceplate more easily if a sheet of paper is inserted between wooden surfaces. The joint is strong enough to hold during turning, yet can be broken easily.

(2) The tool rest may be used parallel to

Figure 108. Using a skew chisel.

the side, or swung around 90° for turning action on the face of the work. If volume of work warrants, obtain a right-angle rest which has two tangs, one for work on the edge and one for the face.

(3) Use same tools and techniques for face turning as for between-center turning. (See fig. 109.) Since most of the cutting is done close to the horizontal center line of work,

tools are used primarily for scraping.

(4) Smooth finished work by holding a strip of sandpaper (3/0 to 4/0) lightly against the work while the lathe is running at medium or low speed. (See fig. 110.) Move sandpaper from side to side and backward and forward. To avoid rounding sharp edges and flattening beads, use narrow strips of sandpaper to smooth these parts.

Figure 109. Faceplate turning.

Figure 110. Sanding and polishing on lathe.

CHAPTER 5

GLUES AND GLUING TECHNIQUE

54. General

Gluing is used extensively in constructing wood parts of unusual form or dimension and in connecting parts of built-up assemblies. The following factors determine the efficiency of glued joints:

  a. Kind of wood.
  b. Moisture content of wood.
  c. Type of joint.
  d. Precision with which contact surfaces match.
  e. Type of glue and method of preparation, handling, and application.
  f. Degree and duration of pressure used in gluing.
  g. Method of conditioning glued joints.
  h. Service conditions.

55. Gluing Properties of Woods

Satisfactory glue joints depend on the density and structure of the wood, the presence of extractives or infiltrated materials in the wood, and the kind of glue used. In general, heavy woods are harder to glue than light woods; hardwoods are harder to glue than softwoods; and heartwoods are harder to glue than sapwoods. Some species vary considerably in gluing properties when different glues are used. Gluing characteristics of common woods are given in table XXIX.

56. Effect of Moisture Content

Wood absorbs moisture from glue and may warp after gluing unless its original moisture content is correct. If stock to be glued is ½ inch thick or more, the moisture content should be from 8 to 12 per cent. For veneers and thin laminations, a moisture content of 5 to 10 per cent is satisfactory. If film glues are used in conjunction with hot pressing, stock should have a moisture content near 10 per cent when

Table XXIX. Classification of various woods according to gluing properties.

Species	Heart-wood*	Sap-wood*	Species	Heart-wood*	Sap-wood*
Alder, red	2	2	Gum, tupelo	4	3
Ash, commercial white	3	3	Hemlock, western	1	1
Aspen	1	1	Hickory, pecan	2	2
Basswood	2	2	Hickory, true	4	4
Beech	4	3	Larch, western	2	2
Birch, sweet and yellow	4	3	Magnolia	3	3
Butternut	2	2	Mahogany	2	....
Cedar, Alaska	2	2	Maple, red and sugar	3	3
Cedar, eastern red	2	2	Oak, commercial white	3	3
Cedar, western red	1	1	Oak, commercial red	3	3
Cherry, black	3	3	Persimmon	....	3
Chestnut	1	1	Pine, northern white	1	1
Cottonwood	2	2	Pine, ponderosa	2	1
Cypress, southern	2	1	Pine, southern yellow	2	2
Douglas fir	2	2	Poplar, yellow	2	2
Elm, American	3	3	Redwood	1	1
Elm, rock	3	3	Spruce	1	1
Fir, commercial white	1	1	Sycamore	3	3
Gum, black	4	3	Walnut, black	3	3
Gum, red	4	3			

*KEY:

1. Woods that glue easily with different glues under a wide range of gluing conditions.

2. Woods that glue satisfactorily with different glues and with moderate care in the gluing operation.

3. Woods that glue satisfactorily if gluing conditions are carefully controlled.

4. Woods that require special treatment before gluing to obtain the best results.

it is put in the press, regardless of its thickness. Moisture content should be uniform throughout the piece.

57. Strength of Glued Joints

a. SIDE-GRAIN JOINTS. Side-grain surfaces can be glued easily. They are stronger than joints made by gluing surfaces cut at an angle to the grain.

b. END-BUTT JOINTS. It is almost impossible to make end-butt joints strong enough for ordinary service. With the most careful gluing possible, not more than about 25 per cent of the tensile strength of wood parallel to the grain can be obtained.

c. SCARF JOINTS. Plain scarf joints are recommended when splicing is necessary. Minimum scarf-joint slopes to insure joint strength approximately equal to wood strength are listed in table XXX.

Table XXX. Minimum scarf-joint slopes

Species	Slope
Birch	1 in 12
Gum, red	1 in 8
Mahogany	1 in 10
Poplar, yellow	1 in 8
Oak, red	1 in 15
Oak, white	1 in 15
Walnut, black	1 in 15

d. GLUE BLOCKS. Glue blocks increase the strength of joints materially. The standard triangular type is satisfactory for box corners; in frame joints, a spline inserted in a narrow groove in the inside edge of the frame members serves as a block. (See fig. 111.)

e. DOWEL JOINTS. The insertion of dowels strengthens glued joints appreciably. This is especially true when end-grain pieces are glued to side-grain pieces, and in end-grain butt joints. Dowels also aid in alignment.

(1) Dowels may be any shape. The conventional type is round and is made of birch or maple. A slight rounding of the edge on both ends of a dowel facilitates assembly.

(2) Dowel holes should be laid out carefully to match the companion holes in pieces to be glued. Drill the holes small enough so dowels will not drop out before glue is applied, and slightly deeper than the length of dowel to be

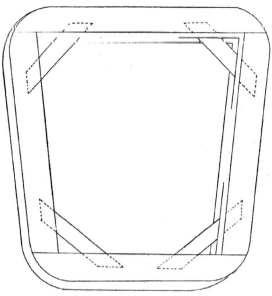

Figure 111. Corner splines.

inserted. This allows for any surplus of glue. Apply thin film of glue in holes rather than to dowels.

58. Contact Surfaces

The strength of a glued joint depends on the care and precision with which contact surfaces are prepared. Although perfect contact is impossible because of minor surface irregularities, the maximum practicable contact area should be secured. To this end, planers, shapers, stickers, saws, and other woodworking machinery should be in such condition that they mark the wood as little as possible. Use plain joints whenever possible because the difficulty of machining or shaping a complicated tongue-and-groove joint cuts down the contact area and nullifies other design advantages. In some cases, however, the tongue-and-groove or splined joint facilitates alignment.

59. Types of Glue

a. FACTORS IN SELECTION. Some of the factors that determine the selection of a glue are—

(1) Ease of preparation.

(2) Working life and assembly time.

(3) Ease of application.

(4) Pressure and heat necessary.

(5) Staining characteristics.

(6) Dulling effect on cutting tools.

(7) Durability.

b. Dry Animal Glues. Dry animal glues are used extensively in furniture and cabinet construction. They are made in dry flake or granular form, are easy to prepare, and have a working life of at least 8 hours after preparation. They are applied either by hand or by a mechanical spreader. Both wood and glue must be warm. The glue begins to set almost immediately; therefore, assembly time is kept under 5 minutes and, if possible, under 3 minutes. The work is cold-pressed and clamps can be removed after 2 hours. Animal glues have practically no staining tendency and only a moderate dulling effect on tools. They have a high dry strength but weaken rapidly when exposed to moisture.

c. Liquid Animal and Fish Glues. Liquid animal and fish glues are used mainly for small jobs and repair work. They are supplied in liquid form ready for use and are applied cold by hand. Assembly time is somewhat longer than for dry animal glues, but should not exceed 5 minutes. The work is cold-pressed for 2 to 3 hours. Like dry animal glues, liquid glues cause little or no staining and dull tool edges only slightly. Better grades have a moderate dry strength, but all grades weaken rapidly when exposed to moisture.

d. Blood-Albumin Glues. Blood-albumin glues are mixed at the time of use and applied cold, either by hand or by a mechanical spreader. Length of working life varies, and the manufacturer's recommendations should be followed. Assembly time is the same as for liquid animal and fish glues. Most blood-albumin glues require hot pressing, which limits their use. They stain wood only slightly and have a moderate dulling effect on tool edges. They are high in strength both when dry and after exposure to moisture.

e. Starch Glues. Starch glues are used occasionally in furniture construction. They are made in powder form. Before application, they are mixed with water and caustic soda and heated. They are applied cold, either by hand or by a mechanical spreader. They have a long working life and a fairly long assembly time. Joints are cold-pressed and left under pressure 4 to 6 hours. These glues stain some woods slightly and have a moderate dulling effect on

tool edges. They have a fairly high dry strength but weaken rapidly when exposed to moisture.

f. Casein and Vegetable-Protein Glues. Casein and vegetable-protein glues are rarely used in furniture work, although they are used extensively in gluing lumber, veneer, millwork, plywood, and in aircraft work. The glues are supplied in powder form. They are mixed with cold water, are usually spread by hand, and are pressed cold. Their working life is 4 hours. Assembly time is not over 20 minutes. Clamping pressure is maintained for at least 2 hours if some heat is used, and for at least 3 hours otherwise. These glues stain some woods badly and have a pronounced dulling effect on tool edges. Joints made with them have a high dry strength and some moisture resistance.

g. Synthetic Resin Glues. The most common and successful types of synthetic resin glues are—

(1) *Urea-resin glue.* Urea-resin glue comes in powder or liquid form. It is mixed with cold water and applied cold by hand or mechanical spreader. Joints are either cold or hot-pressed. For cold-pressed joints, pressure is maintained 4 hours. For hot-pressed joints, pressure time is 5 minutes, after tempeature at the glue line has been raised to about 250°F. The glue stains wood slightly and has a moderate dulling effect on tool edges. Joints made with it have high dry strength. It is more moisture-resistant than casein glue. Cold-setting, urea-resin glue cannot be used if shop or wood temperature is below 70°F.

(2) *Phenolic-resin glues.* Phenolic-resin glues are used in all outside grades of plywood and in exposed gluing assemblies because they have a higher moisture resistance than other glues. They come in aqueous suspension or in dry-film form. In dry-film form, working life and assembly time are indefinite. Hot pressing is essential with both dry-sheet and liquid forms. The glues have a moderate staining effect and a moderate dulling action on tool edges. They are recommended for any assembly in which heat and pressure can be used to set the glue bond. Properly made, the joint is almost indestructible, regardless of exposure conditions.

60. Pressure

All glue joints require pressure to hold the two

wood surfaces in contact while the joint is setting. Pressure is usually applied with some type of screw clamp. (See fig. 112.) If surfaces are well matched and areas are small, little pressure is required. However, pressures up to the crushing strength of the wood may be required for large surfaces. The amount of pressure needed varies also with glue consistency. (See fig. 113.) Pressures must not exceed the crushing strength of the wood. If heat is not used, pressure is maintained for at least 2 hours. When hot-setting, urea- or phenolic-resin glues are used, pressure time can be cut to less than 5 minutes. Figure 114 shows strength-time relationship for casein glues, which are representative of cold-setting glues in common use in furniture manufacture.

61. Conditioning Glued Joints

a. REDUCING MOISTURE CONTENT. When a water-mix glue joint is cold-pressed, moisture content of wood next to the glue line is increased. If moisture in the wood next to the glue joints is not reduced before the work is surfaced, shrinkage from subsequent drying and conditioning causes joints to sink. In thick stock, the general moisture content is not materially increased; but when the assembly is built up of humerous thin laminations, there is an increase in moisture content which must be reduced and equalized to prevent changes in dimensions or alignment. When thick stock is edge-glued, moisture content is reduced by piling the stock on stickers and drying it for 2 to

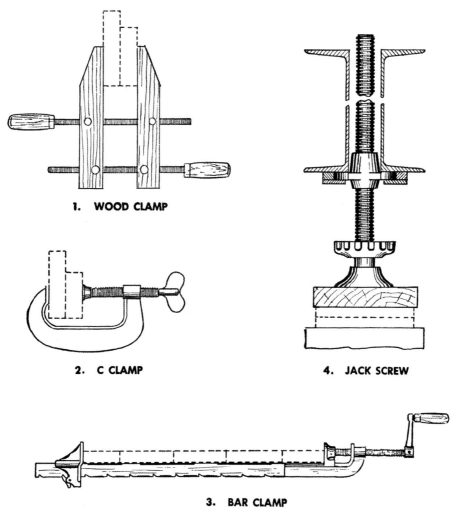

1. **WOOD CLAMP**

2. **C CLAMP**

4. **JACK SCREW**

3. **BAR CLAMP**

Figure 112. Hand devices used in applying pressure to glue joints.

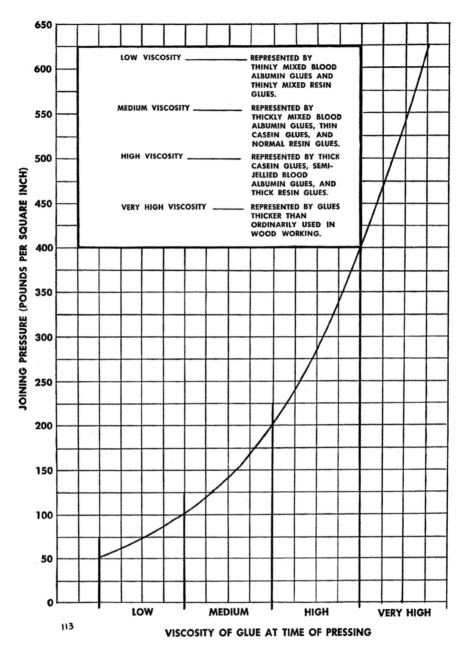

Figure 113. Joining pressures recommended for glue at varying consistencies.

7 days, depending on temperature and air circulation. In thin laminations, moisture is present throughout the wood so the conditioning procedure is similar to that for wet lumber.

b. INCREASING MOISTURE CONTENT. Film glues and hot-press operations tend to dry the wood too much. When this happens, the wood is piled inside a building and moisture content increased by a raising relative humidity to a point where the wood absorbs moisture. Temperature, relative humidity, and air movement are controlled as in kiln-drying. (See par. 5 *c*.)

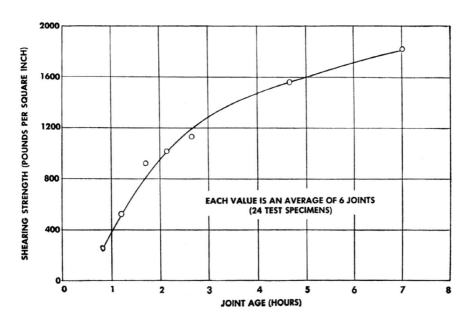

EACH VALUE IS AN AVERAGE OF 6 JOINTS
(24 TEST SPECIMENS)

Figure 114. Strength-time relationship for casein-glue joints.

CHAPTER 6

WOOD FURNITURE

62. General

Major furniture repairs can be largely eliminated by making frequent inspections to detect weak spots, wear, and minor scratches or breaks, and repairing them before they become serious.

63. Solid Surfaces

a. SHALLOW SURFACE DEFECTS. Sand or plane surface to remove shallow defects in solid wood tops.

b. DEEP DEFECTS. If a defect is too deep to be removed by sanding or planing, repair it with shellac filler of a matching color. Clean out the scratch, removing all loose or crushed wood fiber. Enlarge it if necessary and undercut slightly. Apply stick shellac with a hot knife blade, filling the depression to surface level. Smooth the fill.

c. EXTENSIVE DEFECTS. If the surface is so damaged that neither of the above procedures is practical, cover the entire surface with plywood or tempered prestwood cemented down with woodworking glue. First, remove the finish and sand the old surface until it is smooth and free of irregularities. Cut edges of the covering flush with the old top edge. If the old edge is marred, use a thin wood banding of the same finish and species as the original surface. Make sure the top edge of the banding is flush with or very slightly under the surface level of the new top. Make the banding wide enough to cover both old edge and surface material. A tempered prestwood surface need not be finished.

d. WARPED TOPS. Replace warped tops. To increase stability, use a glued-up board instead of a solid one. For instance, if an 18- by 36-inch top is needed, glue up a top from three 6-inch pieces, or rip an 18-inch piece into three 6-inch pieces and reglue. (See fig. 115.)

64. Veneered Surfaces

a. MINOR DEFECTS. Use stick shellac to repair minor defects in veneered surfaces. (For procedure, see par. 63*b*.)

1. SOLID BOARD, SHRINKAGE TENDENCY INDICATED BY DOTTED LINES

2. GLUED-UP CONSTRUCTION BALANCED STRESSES

Figure 115. To prevent shrinkage shown in (1), rip board into thirds, invert center section, and reglue.

b. SMALL DEFECTIVE AREAS. If damage is confined to a small area, repair it as follows:

(1) Select a patch slightly larger than the damaged section. Apply three or four small spots of glue to the damaged area, press the patch over the glue and allow it to set.

(2) With a sharp knife held vertically, cut through both patch and damaged veneer. This cut need not follow a rectangle; it is better to taper it to a point.

(3) Detach the patch and clean out the damaged veneer within the cut area. Apply glue, insert patch, and place a weight on it. Remove excess glue from the surface and allow the repair to set.

c. EXTENSIVE DAMAGE. If the damaged area is too extensive for repair by these methods, cover the entire surface, using the procedure in paragraph 63*c*.

65. Locks and Catches

Minor shrinkage or warpage can make door locks and catches fail to work properly. To correct this condition, shift strike plates or adjust hinge positions.

66. Drawers

a. STICKING. Sticking drawers can be freed by sanding or planing sides or bottom. Apply powdered soapstone to relieve minor sticking.

A slight readjustment of the drawer guides may eliminate a great deal of work on a warped drawer.

b. LOOSE JOINTS. If glue joints at corners show signs of loosening, repair them with glue blocks (par. 57*d*) to prevent dovetail corners from breaking. Do not repair loose corners with brads, because they may split the wood.

67. Fastenings and Attachments

a. ENLARGED SCREW AND NAIL HOLES. Screw or nail holes often become enlarged, with resultant loosening of hinges, strike plates, latches, handles, and similar fittings. Fill such holes with plastic wood or a soft wood plug. Replace the fitting and fasten it securely.

b. WEAR AROUND FITTINGS. If wood surfaces around a fitting become seriously worn, remove the fitting, inlay a new piece of wood in the worn area, and refasten the fitting. If this is not practicable, relocate fitting. Use the following method to relocate butt hinges for doors or lids:

(1) Mortise door or lid to a depth equal to the double hinge thickness. This eliminates the second mortise and reduces chances of an error in marking or mortising.

(2) Fasten hinges on door or lid. Cut off one screw just long enough to project about $\frac{1}{16}$ inch through the hinge when it is closed and the screw is in place. File a point on this stub and set it in the hinge.

(3) Set door or lid in position and press hinge against frame. Drill screw hole in frame at point marked by stub screw.

NOTE. Keep stub-screw marker for future use.

68. Miscellaneous Repairs

a. SPLITS OR CRACKS. Repair lengthwise splits or cracks extending entirely through a member by forcing glue into crack and then applying pressure to close it. Maintain pressure until glue is dry.

b. CROSS-BREAKS. Repair cross-breaks by splicing, or replace entire member. Splicing by scarf joint requires closely fitting contact surfaces. Hand tools can be used with a simple jig, such as the one shown in figure 116, to insure accurate work. Using Jorgensen or C-clamps, fasten jig, guide, and pieces to be spliced to a work bench, and cut bevel by slicing a hand plane, side down, against edge of the jig base.

c. BROKEN MORTISE JOINTS. To repair broken mortise joints, butt-glue broken ends together. Reinforce joint with a screw or dowel long enough to penetrate at least 1 inch into tenon member. Do not use this method to repair highly stressed members; replace such members completely.

d. LOOSE DOWELS. Dowels sometimes shrink and become loose at one end, allowing the joint to open. If the dowel cannot be replaced, repair it by one of the following methods:

(1) Make a saw cut lengthwise through

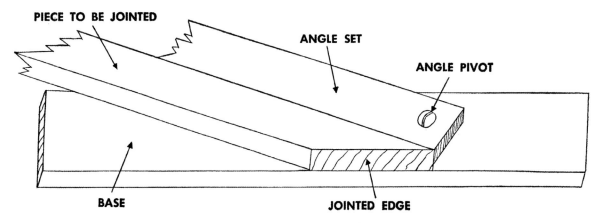

PIECE TO BE JOINTED

ANGLE SET

ANGLE PIVOT

BASE

JOINTED EDGE

Figure 116. Jig for scarf joint.

shrunken end of dowel. Insert small wedge in cut, with wide end of wedge projecting beyond end of dowel, and force dowel back into joint. (See fig. 117.) This drives wedge deeper into dowel, expanding it and making joint tight.

(2) If wedging is not practicable because member cannot be disassembled, anchor loose dowel end in the joint with a smaller cross dowel or screw.

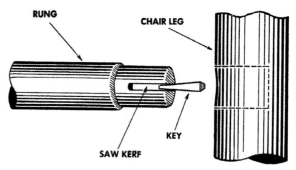

Figure 117. Wedging a dowel joint

CHAPTER 7

UPHOLSTERED FURNITURE

69. Repair Problems

Upholstered furniture consists basically of a frame, strip or cleat webbing, padding, and cover. Some furniture also has spring upholstery. Repairs needed on upholstered furniture generally include recovering, replacement or redistribution of padding, replacement or refastening of webbing, and regluing, reinforcing, or replacing frame parts. With spring construction, replacing, anchoring, and retying springs may also be necessary.

70. Recovering

a. GENERAL. Replace the entire cover if the covers on seat, back, or arms are torn, soiled, or worn beyond repair. Even with fairly new furniture, it is usually impossible to match new material to worn or faded fabric, so all sections must usually be recovered when one is damaged.

b. PROCEDURE. Procedures for recovering upholstered furniture vary with furniture design, but the following general procedure applies to almost all types:

(1) Remove old cover carefully, taking out all tacks.

(2) Using the old cover as a pattern, cut a piece of new material to approximate shape and size.

(3) Smooth out and replace any lumpy or torn padding and lay new cover in place, making certain all four sides have the same amount of surplus material.

(4) Tack center of opposite sides, stretching the material lightly but firmly. Do not drive tacks all the way in. Work from center to edges, stretching material evenly. If wrinkles develop, remove tacks and work the wrinkles out. Note how the old covering was folded and fitted at corners and around legs and arms. If this was satisfactory, fit the new cover the same way. When covering fits smoothly, drive tacks all the way in.

(5) After covering is tacked to the side of the frame, cover tack heads with an edging or gimp. Fasten gimp with large-headed upholstery nails spaced about 2 inches apart.

> NOTE. Whenever covering and padding are removed, overhaul frame, webbing, and springs. (See par. 72, 73, and 74.)

71. Replacement or Redistribution of Padding

a. GENERAL. Padding of tow, cotton batting, excelsior, or moss is used over the springs in the case of spring construction, or on the webbing in the case of padded construction. When padding shifts or becomes lumpy, remove the cover and redistribute or replace the padding.

b. PROCEDURE. To replace padding—

(1) First, remove all old padding and tack a piece of burlap smoothly over the entire surface to be padded.

(2) Spread padding evenly over the burlap, forming a compact cushion about 1½ inches thick.

(3) Cover this with a second piece of burlap or muslin, tacked down securely, and place a 2-inch layer of cotton batting on top. Pull off surplus cotton around the edges; *do not cut* the cotton since this will make a ridge under the cover.

(4) Tack a cambric cover over the frame bottom to keep padding from working through to springs or webbing and falling out.

(5) Replace cover as described in paragraph 70.

72. Repairing Frame

Tighten loose frame joints with glue blocks, pins and screws, or angle irons. Repair other frame damage using procedures in chapter 6.

73. Repairing Webbing

Check strip or cleat webbing for signs of wear or breakage whenever cover is removed. Replace damaged webbing and refasten loose strips. To insure that webbing will hold securely, double it over at the ends to give tacks more gripping power, tighten, and tack so stress is at right angles to tack length. Run webbing in two directions, at right angles to each other. (See fig. 118.) Closing the entire bottom with webbing is not necessary, but too much webbing is better than too little. If springs are to be anchored to webbing, space the webbing to support spring bases. Similarly, anchor metal strips or wood cleats securely and space them for springs.

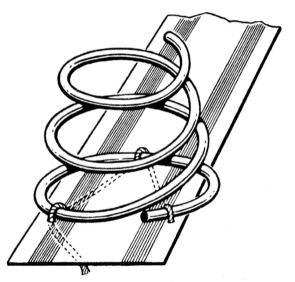

Figure 119. Tying spring to webbing.

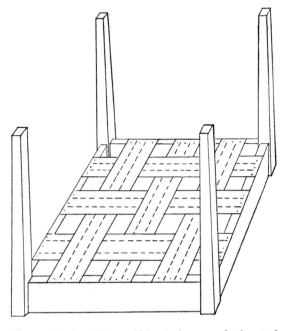

Figure 118. Attaching webbing to framework of a stool.

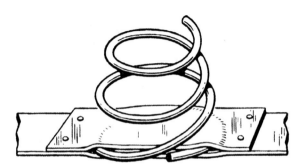

Figure 120. Fastening spring to metal strip.

74. Adjusting Springs

Springs may shift, bend, or become damaged otherwise. Re-anchor and retie loose springs; replace those that are damaged.

a. ANCHORING. Springs are usually attached differently on webbing, on metal strips, or on wood cleats. Fasten spring bases to webbing with heavy flax cord about 1/8 inch in diameter. (See fig. 119.) Anchor springs to metal strips with clamps. If clamps loosen, rerivet them. (See fig. 120.) Fasten springs to wood cleats with staples or metal straps and nails. (See fig. 121.)

b. RETYING. After springs are anchored, retie them with heavy flax cord like that used to anchor springs to webbing.

(1) Nail cord to the center of one side of the frame. Pull it over the top of the springs

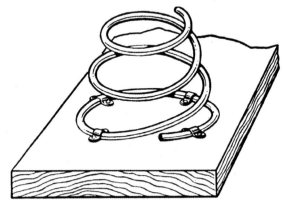

Figure 121. Fastening spring to wood cleat.

to an opposite anchoring nail. Allow enough cord to tie two double half hitches to each spring and cut to this length.

(2) Bring cord up to the top of the first

coil spring and tie it with a double half hitch to the nearest rim of the first spring. Before drawing the knot tight, pull spring down to shape the seat or back. (See fig. 122.) Continue to opposite side of the top on the same spring and tie it.

(3) Continue in like fashion, tying two points on each spring and finally anchoring cord to nail on opposite side of frame. Run cords in both directions (side to side and front to back) at right angles to each other, until all springs are tied in two directions.

(4) Tie springs diagonally in the same manner, beginning at one corner of the frame and anchoring cord on the opposite corner.

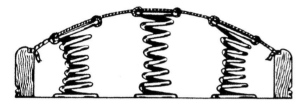

Figure 122. Springs tied in place.

(5) Repeat with cord at right angles to the first set of diagonal cords. Tie this cord to spring with two double half hitches and also tie it to the other three cords at their junction in the center of the coil. Each spring is now

tied in eight places and the crossing cords are also tied together. Completely retied springs are shown in figure 123.

(6) Replace padding. (See par. 71.)

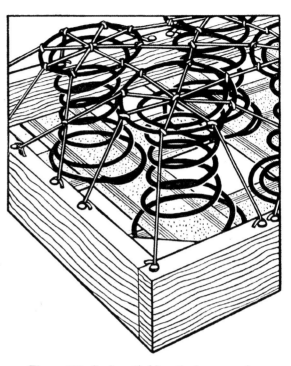

Figure 123. Springs tied lengthwise, crosswise, and diagonally.

CHAPTER 8

REFINISHING

75. Standards

Finishes should be kept sufficiently neat and attractive in appearance to give the users reasonable satisfaction, and so encourage careful use and maintenance of the furniture. Badly scarred or worn finishes tend to promote careless treatment and further damage. On the other hand, unnecessarily elaborate finishes are inappropriate. Finishes should be adequate for easy cleaning of surfaces and for protection against rapid absorption of moisture. As a rule, the standards of commercial furniture should be approximated.

For precautions in the use of toxic or inflammable materials, see TM 5-618 (when published).

76. Types of Finish

Furniture can be finished with varnish, lacquer, paint, or enamel. Repair the old finish with new material of the same type. If a different finish is desired, remove all of the old finish and refinish completely.

a. VARNISH. Varnish finishes show the grain and texture of the wood. The wood may be left in its natural color or stained before varnish is applied. When using an oil stain, apply a coat of wood sealer before applying varnish or lacquer. To obtain smooth surfaces, use wood filler before varnishing hardwoods with pores larger than those in birch. Varnish dries more slowly than lacquer, but is usually easier to apply.

b. LACQUER. Usually, two coats of lacquer are required for a coating as thick as one coat of varnish. Since lacquers are economical for large-scale factory production, much of the furniture supplied under Quartermaster Corps specification 307-E is finished with lacquer.

c. ENAMEL. Enamel finishes are opaque and provide smooth, hard surfaces. Use them only when surfaces have been carefully prepared and when large pores have been filled with wood filler. Apply enamel undercoater and sand it smooth before enameling. Oleoresinous enamels resemble varnishes in drying time, number of coats required, and convenience for small-scale work; lacquer enamels resemble lacquers.

d. PAINT. Paint finishes hide surfaces, but do not form perfectly smooth coatings. First, use a suitable primer, then one or two coats of paint. In general, paint finish is used on less expensive furniture where enamel finish is not required.

e. OTHER FINISHES. Concealed wood surfaces, such as the undersides of table tops, backs of side panels, and interiors of drawers, need protective finishes to retard changes in moisture content. Approximately the same protection is required as on exposed surfaces, to minimize the tendency to warp. Use two coats to insure satisfactory protection, because one coat rarely forms a continuous covering. Suitable protective finishes are aluminum paint, gloss paint, shellac, varnish, or wood sealer. Linseed oil furnishes little protection against moisture changes.

77. Repairing Old Finish

Whenever practicable, the old finish should be repaired rather than replaced. This is usually possible in the following cases:

a. AFTER REPAIR OF INTERIOR MEMBERS. After repairing interior members of furniture, remove all excess glue from exposed surfaces before it hardens. Repair minor scars in the finish with stick shellac (par. 63*b*) and furniture polish or furniture wax.

b. DULLED FINISH. If finish is intact but has become dull from accumulated dirt, wash surfaces with mild soap and warm water, rinse quickly with warm water, and then rub with furniture polish or wax. If a varnish finish has become dull from age or exposure to sunshine or has been worn thin, wash, dry, recoat with varnish, and rub with FFF pumice and crude oil. To repair paint and enamel finishes without removing the old finish, wash or sandpaper the surface lightly and recoat it with paint or enamel.

c. WORN SPOTS. When finish is worn in a few small spots, touch up bare spots and apply a coat of finishing material to the entire surface. (See also par. 78*c*.) If there are many worn spots or if their area is great, it is usually more economical to remove all old finish.

78. Surface Preparation

Proper surface preparation is essential for a satisfactory finish. All surfaces should be smooth and free from dirt, dust, and grease.

a. PAINT FINISHES. Paint finishes are easiest to match and maintain. Apply priming coat under finish coat on exposed or new wood. Cover knots with a coat of shellac or, preferably, aluminum paint. One or two coats of paint following a priming coat are usually enough for a good solid color. Aluminum paint is an excellent priming coat.

b. ENAMEL FINISHES. To repair enamel finishes, sandpaper any areas of bare wood until edges of the old coating are feathered smoothly. If wood has pores larger than those in birch, apply natural wood filler. Touch up all bare areas with enamel undercoater and sand with 5/0 and 8/0 sandpaper when dry. Repeat until surface is smooth, then apply one or two coats of enamel over entire surface.

c. STAINED AND NATURAL FINISHES. Stained and natural finishes are hardest to match since new wood rarely matches the color of old wood. Oil stains and water stains are commonly used on new furniture; spirit (alcohol) stains are often used in repair work. For natural finish, color new wood slightly with oil stain made from painters' tinting colors. For stained finishes, stain bare areas to match the old wood, using the same stain as in the original finish. After applying water stain, sand the wood lightly. On wood with pores larger than those in birch, apply wood filler after staining. Use filler of the same color as in the old finish. Color natural filler by adding oil stain or painters' tinting colors. Touch up newly stained areas with wood sealer, then apply varnish or lacquer over the entire surface.

79. Methods of Application

a. GENERAL. (1) *Stains.* Apply stains by brushing, dipping, spraying, or in repair work by mopping with a clean rag or cotton waste. Allow oil stains to remain on the surface for a few minutes before wiping off excess. Staining should be done quickly so all laps are picked up while their edges are still wet.

(2) *Wood Filler.* Apply wood filler with a brush and allow it to stand a few minutes. Then wipe off excess with burlap, clean rags, or cotton waste, stroking first across the grain of the wood and then lightly along the grain.

(3) *Sealer.* Apply sealer by spraying, brushing, or dipping. Spraying gives the best results.

(4) *Varnish, paint, and enamel.* Apply varnish, paint, oleoresinous enamel, paint primer, and enamel undercoater, either with a brush or spray gun.

(5) *Lacquer.* Apply lacquer and lacquer enamel by spraying, except on very small areas.

b. SPRAYING. (1) *Procedure.* When a spray gun is used to apply a finish.

(a) Make sure equipment is in good condition. See that nozzle jets or openings are clear, that there is no leakage in the system, and that there is enough pressure. Check fluid tip and air-cap combination to make sure they are suitable for the finish being applied.

(b) If paint is being applied, stir it thoroughly and thin it to the consistency recommended by the spray-gun manufacturer. If paint contains lumps or skins, strain it through a screen.

(c) Hold the spray gun perpendicular to surface and about 6 to 10 inches away. Move spray with a free arm motion. (See fig. 124.) Feather the ends of the strokes by triggering the gun; that is, by starting the stroke before pulling the trigger and by releasing the trigger just before ending the stroke. Spray within 1 or 2 inches of corners and then spray both sides of the corner at once. (See fig. 125.)

(2) *Spraying defects.* (a) *Defective patterns.* Defective patterns are caused by clogging of one more horn holes or by dirt on the fluid tip or air-cap seat. Heavy center patterns are usually caused by setting the spreader adjustment valve too low or, when twin-jet cap is used, by low atomizing pressure or to material with too great viscosity. When pressure feed is used, center patterns are caused by a fluid pressure too high for the capacity of the cap or by a nozzle too large for the material used. Split-spray pattern is caused by improper bal-

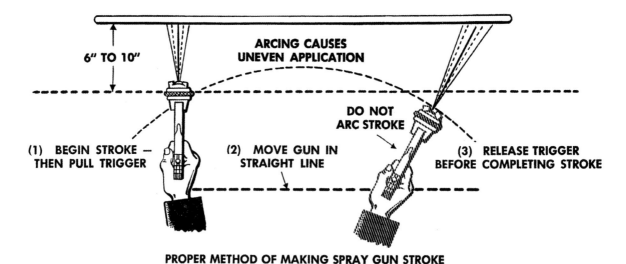

ARCING CAUSES UNEVEN APPLICATION

6" TO 10"

DO NOT ARC STROKE

(1) BEGIN STROKE — THEN PULL TRIGGER

(2) MOVE GUN IN STRAIGHT LINE

(3) RELEASE TRIGGER BEFORE COMPLETING STROKE

PROPER METHOD OF MAKING SPRAY GUN STROKE

Figure 124. Using spray gun.

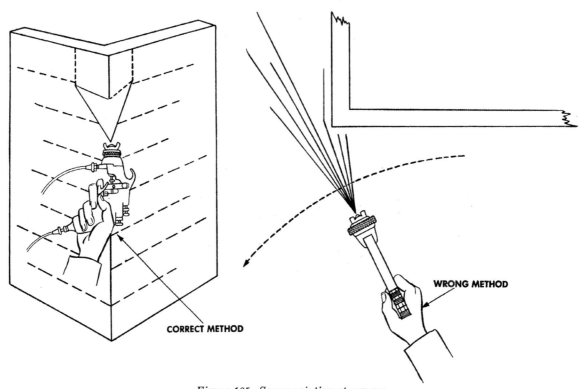

CORRECT METHOD

WRONG METHOD

Figure 125. Spray painting at corners.

ance of air and fluid. To correct this, reduce width of the spray pattern with spreader adjustment valve or increase the fluid pressure.

(b) Orange-peel finish. A common cause of orange-peel finish is use of a thinner with a high percentage of low-boiling solvents. Other causes are insufficient atomization, holding the gun at the wrong distance from the surface, not dissolving or stirring the material thoroughly, drafts in finishing room when applying synthetics and lacquers, or too low humidity when spraying synthetics.

(c) *Streaks.* Streaks are caused by tipping the gun so one side of the pattern hits the surface from a shorter distance. The uneven patterns described in *(a)* above also give this effect.

(d) *Runs and sags.* Runs and sags are the result of applying too much finish on the surface. To remedy this, cut down fluid pressure or increase operating pressure. Unequal patterns also contribute to runs and sags.

The following chart shows a typical blush curve for a representative lacquer of good brush characteristics. Any condition of temperature and relative humidity within the area above the curve will produce blushing. Blushing will not occur under conditions below the curve.

Several temperature and relative humidity readings daily over a period of time, along with notations of the current behavior of the film, will provide data which will permit plotting an exact curve of blush probability for a lacquer applied in a particular finishing department.

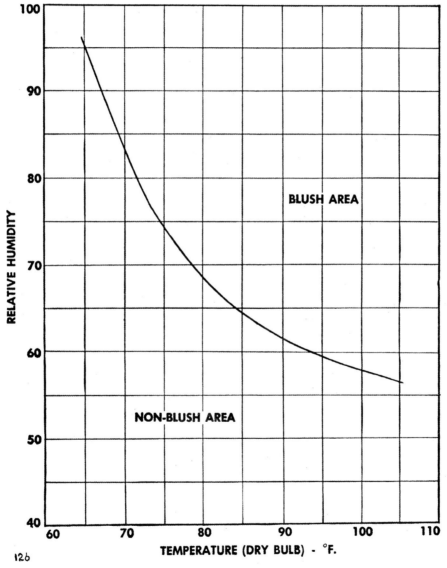

126

Figure 126. Blushing chart.

(e) Misting and fogging. Excessive misting or fogging results from overatomization, caused by too high atomizing air pressure, wrong air cap, wrong fluid tip for material used, low fluid pressure, or incorrect distance of the gun from the surface.

(f) Blushing. Blushing is caused by moisture in lacquer. Even though all moisture has been eliminated from the lacquer and the air system, blushing can occur because of absorption of moisture from the air (fig. 126) or by condensation of moisture as a result of difference between temperature of the article and the surrounding air. To avoid this, control relative humidity and, if necessary, keep the article to be sprayed in warm storage.

80. Finishing Schedule

a. MATERIALS. Finish officers' and noncommissioned officers' furniture with the following materials:

(1) Walnut transparent water stain for officers' furniture; mahogany for noncommissioned officers' furniture.

(2) Du Pont's No. V 7782, Sherwin-Williams' No. 01954, Cromar Company Tone Phlex, or equal quality wood primer.

(3) Silex-paste wood filler which can be mixed with turpentine or benzine and can be colored to shade of water stain.

(4) Du Pon't No. 1935, Sherwin-Williams' No. 89 Opex, or equal quality sealer.

(5) Du Pont's No. 1734, Sherwin-Williams' No. 674 Clear Opex, or equal quality clear gloss lacquer.

(6) Du Pont's No. 1737, Sherwin-Williams' No. 92 Opex, or equal quality flat lacquer.

(7) Du Pont's V W 6757, Sherwin-Williams' No. 28529 Opex Furniture Cleaner, or equal quality lacquer cleaner.

(8) Du Pont's No. 231451 White Pyroxylin Surfacer, Sherwin-Williams' No. 29177 Opex White Surfacer, or equal quality surfacer for kitchen tables, except tops.

(9) Du Pont's No. 2371 White Duco, Sherwin-Williams' No. 30527 Opex White, or equal quality enamel for kitchen tables, except tops.

b. HOUSEHOLD AND OFFICE FURNITURE. The finishing schedule below, which is specified for new furniture procured under government contract, is a guide to refinishing furniture.

(1) *Stain.* Treat all exposed surfaces and backs of cases and mirror frames with one coat of transparent water stain. Mix stain with hot water and apply it after the mixture has cooled. When finish is dry, sand the piece lightly.

(2) *Wood sealer.* Dip or spray all pieces except kitchen tables with one coat of Du Pont No. V 7782 or equivalent. Wipe off excess material within 10 minutes after dipping to avoid sags. Air-dry at least 2 hours in a well-ventilated space. Remove all drawers during drying.

(3) *Filler.* Mix silex-paste wood filler with turpentine or benzine in ratio of 12 pounds of filler to 1 gallon of reducer. Color with oil stain or painter's tinting colors. Stain in filler may have to be darker than usual because it does not penetrate as easily as water stain. Apply filler by brush, stroking across grain until the pores are filled. Fill pores level with the wood surface. Wipe away all excess filler. Allow article to dry for a short time and then recheck for poorly filled areas. If necessary, apply more filler, rubbing with a circular motion over the entire surface. Wipe clean a second time and dry for 24 hours. Sand lightly.

(4) *Sealer.* Apply sealer coats to all exterior surfaces, dry thoroughly, and sand before applying lacquer. Do not use shellac for this operation.

(5) *Lacquer.* Apply three medium coats of clear lacquer, with not more than 24 hours between coats. Air-dry the final film 36 hours before rubbing or polishing. For rubbing, use FFF pumice and crude oil. Rub until surface is smooth, transparent, and semiglossy. Clean article with a soft rag, then use lacquer cleaner for final cleaning.

c. KITCHEN FURNITURE. Treat tops of kitchen tables with a coat of linseed oil diluted with equal parts of toluol or turpentine to insure good penetration. This provides a surface that wears well and can be cleaned easily. Refinish all kitchen furniture except table tops as follows:

(1) Apply two coats of Du Pont's No. 2314591 White Pyroxylin Surfacer, or equal. Sand after each coat.

(2) Apply two coats of Du Pont's No. 2371 White Duco, or equal. Allow 24 hours between coats.

CHAPTER 9

METAL FURNITURE

81. Repairs

Damage to metal furniture usually consists of distortion, dents or bulges, cracks or tears, or distinct cross-breaks.

a. RESTORING. (1) *Distortions.* Straighten distortions by heating the bent member with a welding torch or a blowtorch and working it back into correct alignment.

(2) *Breaks.* Close cracks or tears by running a bead of weld along the length of the separation. Treat cross-breaks the same, but reinforce by tack-welding a 0.065 metal patch to the member above and below the break.

b. BRACING. Reinforce severely stressed members with angles or U-braces. Tack-weld braces in angles or corners so they do not detract from the appearance of the furniture.

c. REFINISHING. When heat is used to repair metal furniture, the adjacent surface usually must be refinished. Refinishing an entire surface or assembly is usually better than blending in a small touch-up job. Sand the old paint down to the metal and file all welds smooth. Clean the surface with a metal conditioner, then coat it with metal primer such as zinc chromate. Finish with three to five coats of enamel or lacquer, lightly sanded between coats to produce a deep lustre.

APPENDIX I

DEFINITIONS

Accretion. Growth (of trees) by the addition of layers of tissue.

Angiosperm. A plant having seeds inclosed in a vessel called a pericarp (hardwoods).

Astragal. A convex molding or bead.

Bolt. A block of timber to be sawed or cut into pieces such as staves or shingles; a short, round section of a log.

Burl. A wartlike growth on a tree trunk containing piths of many undeveloped buds. These form attractive patterns when the trunk is cut into lumber.

Cellulose. A substance constituting the chief part of the solid framework (cell walls) of plants.

Creosote. Preservative used to prevent growth of fungus in lumber.

Conifer. Plant which bears cones (softwoods).

Deciduous. A plant which sheds leaves annually (hardwoods).

Dicotyledon. A plant whose seeds divide into two lobes in germinating (hardwoods).

Equilibrium moisture content. That moisture content of wood at which no expansion or contraction takes place (providing the relative humidity of air remains constant).

Fiber saturation. Water filling only the wood substance in the lumber and not the cavities.

Free water. Water in excess of that held in the wood substance; also, that which is held in the cavities between cells.

Fungus. Microscopic plant growth feeding on wood and causing stain, mold, or rot.

Gain. A mortise or notch cut to receive a hinge or other fitting.

Gymnosperm. A plant having seeds borne upon the surface of the cone scale (softwoods).

Homogeneous. Of the same composition or structure throughout.

Impregnated. Saturated with another substance (wood treated to protect it from decay).

Kiln-drying. Method of artificially seasoning lumber by heat and air circulation.

Lignin (lignified). A bonding and stiffening substance, which, with cellulose, forms the main substance of wood (woody).

Lumen. The cavity in a cell.

Miscible. Capable of being mixed.

Microscopic. So small or fine as to be visible only with the aid of a lens.

Monocotyledon. A plant having single seed leaf or cotyledon (palms).

Opaque. Not transparent; impervious to light.

Radial. In the same direction as the radius.

Rays. Strips of cells running in a radial direction; that is, running out from the center.

Saprophytic. Pertaining to an organism that lives on dead organic matter.

Scarf Joint. A diagonally lapped joint, forming a continuous piece.

Species. A classification of a group of plants.

Specific Gravity. Weight of some substance as compared with the standard. A cubic foot of wood substance compared with an equal volume of water is 1.54. (Not a cubic foot of lumber).

Spermatophyta. A seed-bearing plant.

Stability of Wood. The tendency of some woods to remain fixed and not shrink, or swell, or warp.

Stickers. Wood strips used to separate courses in lumber piles.

To tail a machine. To assist the operator of a machine in handling stock that is being processed.

Tangential boards. Lumber cut approximately perpendicular to the radius of a log.

Tensile strength. Strength in tension; resistance to being pulled apart.

Viscosity. Stickiness; gumminess.

APPENDIX II

REFERENCES

1 1. DeVilbiss Company, Toledo, Ohio: The ABC of Spray Painting Equipment (1940).

1 2. DeWalt Products Corporation, Lancaster, Pa.: Instruction Book, GP Model (1943).

1 3. Oliver Machinery Company, Grand Rapids, Mich.: Installation, Care, and Operation of Oliver Circular Saw Benches.

4. PLUMLEY, STUART: Oxy-Acetylene Welding and Cutting, 3rd Edition; University Printing Company, Minneapolis, Minn. (1939).

5. Quartermaster Corps Tentative Specification 307-E (1937, revised 1940). Office of the Quartermaster General, Washington, D. C.

6. SMITH, R. E.: Machine Woodworking; McCormack Mathers Publishing Company, Wichita, Kan. (1938).

2 7. U. S. Department of Agriculture Bulletin 1500: The Gluing of Wood (1929).

2 8. U. S. Department of Agriculture Bulletin 1136: Kiln Drying Handbook; (1923, revised 1929).

2 9. U. S. Department of Agriculture: Wood Handbook, Forest Products Laboratory (1940).

3 10. U. S. Forest Service Mimeo 1340: Control of Conditions in Gluing (1941).

11. War Department TM 9-2852: Welding Theory and Application.

12. War Department TM 10-455: The Body Finisher, Woodworker, Upholsterer, Painter, and Glassworker.

13. WELLS, P. A., and HOOPER, J.: Modern Cabinet Work, 5th Edition; Lippincott (1938).

1 14. Yates-American Machine Company, Beloit, Wis.: Vocational Woodworking Equipment; Catalog 25-S (1930).

1. Available from the manufacturer.

2. Available from the Superintendent of Documents, Washington, D. C.

3. Available from Forest Service, U. S. Department of Agriculture, Washington, D. C.

INDEX

	Paragraph	Page
ADJUSTMENTS, CLAMPS AND STOPS. (*See* Mortiser, hollow-chisel.)		
AIR DRYING. (*See* Lumber, seasoning.)		
ANGIOSPERMS	3*a*(2)	2
ANGLE CUT-OFF with		
Overarm saw	27	75
Ripsaw	15*d*	59
ANIMAL GLUE. (*See* Glue.)		
BAMBOOS	3*a*(2)	2
BAND SAW	47, 48	96
Operation:		
Bevel ripping	48*c*	99
Cutting	48*a*	96
Resawing	48*b*	99
Parts:		
Blade	47a	96
Table	47*b*	96
BEADING With wood shaper	44, 45*c*	93, 95
BETA NAPHTHOL, Use for preservation of wood	7	7
BEVELING With—		
Band saw	48*c*	99
Overarm saw	29	75
Ripsaw	15*d*	59
Single-arbor variety saw	12*d*	54
Wood jointer	40	86
BIRD'S-EYE MAPLE. (*See* Maple, bird's-eye.)		
BLACK GUM, DRYING. (*See* Gum, special drying.)		
BLOCKS:		
Clearance	21*a*(5)(*c*)	72
Stop	21*a*(5)(*b*)	71
BLOOD-ALBUMIN GLUE. (*See* Glue.)		
BURLAP. Use in upholstering furniture	71*b*	117
BURLS	3*d*, 3*e*	2
BUTT WOOD	3*e*	2
CALIPERS AND DIVIDERS. (*See* Tools.)		
CAMBRIC, Use in upholstering furniture	71*b*(4)	117
CASEIN GLUE. (*See* Glue.)		
CELLULOSE	3*f*	3
CHAMFERING With—		
Ripsaw	15*d*	58
Wood jointer	36, 40	84, 86

CHECKS. (*See* Lumber, defects.)

CHEMICAL SEASONING. (*See* Lumber, seasoning.)

CHISELS. (*See* Wood lathe, tools.)

CIRCULAR SAWS ....12-22 incl.   54

    Blades:
        Crosscut 12c, 13b, 16, 17, 19, 20   54, 60, 63, 67, 69
        Dado head 18, 21   64, 70
        Miter 13c, 17   55, 63
        Ripsaw 12c, 13a, 15, 20   54, 58, 69

    Fixtures and jigs...22   74

    Kinds:
        Combination 20a, 21   69, 70
        Cut-off 12a, 13b   54
        Ripsaw 12b, 13a, 15, 20a, 21   54, 58, 69, 70
        Single-arbor variety 12d, 13c   54, 55
        Universal double-arbor 12c   54

    Operation:
        Angle cut-off 15d   59
        Beveling 12d, 15d   54, 59
        Chamfering 15d   59
        Cheek cuts 21c, 21d   72, 73
        Compound miter cuts...17c   64
        Crosscutting 12c, 12d, 13b, 13c   54, 55
        Cutting—
            Concave curves and moldings.....................19a(1), 19c   67, 68
            Contours 19   67
            Convex curves and moldings....................19a(2), 19b   67
            Dadoes 12d, 16d, 18c, 30   54, 63, 67, 75
            Gains 12d, 16d, 18c, 30   54, 63, 67, 75
            Long stock 16b   62
            Moldings 12d, 19   54, 67
            Wedges 20a   69
            Wide stock 16c   62
        Edging 12b   54
        Grooving 12d, 15c   54, 59
        Jointing 12b   54
        Mitering 12d, 17a, 17b   54, 63
        Rabbeting 15b   58
        Ripping 12b, 12c, 15a   54, 58
        Splined miter joints...17d   64
        Squaring ends 16a   60
        Tapering 20b   70

CLASSIFICATION OF TREES. (*See* Trees.)

COMPOUND MITER CUTS With—
    Crosscut saws 17c   64
    Overarm saws 28   75

COMPRESSION FAILURE. (*See* Lumber, defects.)

CONCAVE CURVES AND MOULDINGS. With circular saws..................19a(1), 19c   67, 68

	Paragraph	*Page*
CONIFERS	3a(1)	2
CONTOURS, CUTTING With circular saws	19	67
CONVEX CURVES AND MOLDINGS With circular saws	19a(2), 19b	67
COTTON BATTING, Use in padding of upholstered furniture	71	117
CREOSOTE, Use for preservation of wood	7	7
CROSSCUTTING WITH—		
Circular saws	13b	54
Cut-off saws	12a	54
Miter saws	13c	55
Overarm saws	24	75
Single-arbor variety saws	12d	54
Universal double-arbor saws	12c	54
CUT-OFF SAWS. (*See* Circular saws.)		
CUTTER:		
FLAT-KNIFE. (*See* Wood shaper, parts.)		
SOLID-HEAD. (*See* Wood shaper, parts.)		
DADO HEAD:		
Setting up	18a, 23	67, 75
Used for making:		
Cheek cuts	21d	73
Dadoes and gains	18c	67
Grooves and rabbets	18b	67
Right-angle cuts	18c	67
DADOES, CUTTING With—		
Crosscut saws	16d	63
Dado heads	18c	67
Overarm saws	23, 30	75
Single-arbor variety saws	12d	54
DICOTYLEDONS	3a(2)	2
DIVIDERS. (*See* Tools.)		
DOUGLAS FIR TIMBERS, SEASONING	5d	6
DRAWERS, WOOD FURNITURE:		
Correction of sticking	66a	114
Repair of loose joints	66b	115
EDGE OR VERTICAL GRAIN LUMBER	4a(2), 4b(2)	4, 5
EDGING With—		
Ripsaw	12b	54
Wood jointer	36, 38	84
EXCELSIOR, Use in padding of upholstered furniture	71	117
EXPANSION OF WOOD	5a(1), 5a(2)	5
FABRICS	2a(2)	1
FACE MITERING. (*See* Mitering.)		
FACE-PLATE TURNING. (*See* Wood lathe, turning procedure.)		

	Paragraph	Page
FASTENINGS IN—		
FURNITURE CONSTRUCTION	10	49
WOOD FURNITURE, REPAIR	67	115
FEATHER BOARD, Use in ripping at angle	15d(1)(b)	60
FENCE RIPPING	21a(5)(a)	71
FINISHES FOR WOOD FURNITURE		
Application of finish	79	121
Finishing schedule	80	124
Kinds of finish:		
Enamel	76c	120
Lacquer	76b	120
Paint	76d	120
Protective	76e	120
Varnish	76a	120
Repairing old finishes	77	120
Standards	75	120
Surface preparation (see also Wood furniture, repair)	78	121
FISH GLUE. (See Glue.)		
FITTINGS ON WOOD FURNITURE, REPAIR	67	115
FLAT-GRAIN LUMBER	4a(1), 4b(1)	4, 5
FLUTING With wood shaper	44, 45c	93, 95
FRAME DETAIL IN FURNITURE CONSTRUCTION	9	10
FRAMES FOR UPHOLSTERED FURNITURE, REPAIR. (See Upholstered furniture, repair.)		
FUNGUS	3b(2)	2
FURNITURE CONSTRUCTION	9, 10	10, 49
FURNITURE REFINISHING	75-80 incl.	120
Application of finish	79	121
Finishing schedule	80	124
Repairing old finish	77	120
Standards	75	120
Surface preparation	78	121
Types	76	120
(See also metal furniture, repair)		
FURNITURE REPAIR. (See Metal furniture, repair), (see also Wood furniture, repair; Upholstered furniture, repair)		
FURNITURE WOODS (See also Table I)	8	7
GAINS, CUTTING With—		
Crosscut saws	16d	63
Dado heads	18c	67
GLUE, kinds:		
Animal	59b, 59c	110
Blood albumin	59d	110
Casein	59f	110
Fish	59c	110

	Paragraph	_Page_
Starch	59*e*	110
Resin	59*g*	110
Vegetable protein	59*f*	110
GLUE BLOCKS	57*d*, 66*b*	109, 115
GLUING	54-61 incl.	108
Conditioning glued joints	61	111
Contact surfaces	58	109
Effect of moisture content	56	108
Pressure used	60	110
Properties of wood (_see also_ table XXIX)	55	108
Service conditions	54*h*	108
Strength of joints	57	109
Types of glue	59	109
GOUGES FOR WOOD LATHE. (_See_ Wood lathe, tools.)		
GRAIN DEVIATION. (_See_ Lumber, defects.)		
GRAVITY, LOW SPECIFIC	3*d*	2
GROOVING With—		
Dado head	18*b*	67
Ripsaw	15*c*	59
Single-arbor variety saw	12*d*	54
Special heads	18*d*	67
Wood shaper	44	93
GUM, SPECIAL DRYING	5*b*(2)	6
GYMNOSPERMS	3*a*(1)	2
HARDWOODS	3*a*(2), 3*a*(3), 4*a*, 4*b*, 43*c*, 55	2, 4, 5, 91, 108
HEADSTOCK OF WOOD LATHE. (_See_ Wood lathe, parts.)		
HEARTWOOD	3*b*(2), 55	2, 108
HINGES ON WOOD FURNITURE, REPAIR	67	**115**
HOLLOW-CHISEL MORTISER. (_See_ Mortiser, hollow-chisel.)		
HOPPER CUTS. (_See_ Compound miter cuts.)		
INSECT DAMAGES. (_See_ Lumber, defects.)		
JIGS, SPECIAL	22, 45*c*	74, 95
JOINTER, WOOD. (_See_ Wood jointer.)		
JOINTING With—		
Jointer	36, 37, 38	84, 85
Ripsaw	12*b*	54
JOINTS IN FURNITURE CONSTRUCTION	10	49
JOINTS:		
GLUING	57	109
Dowel	57*e*	109
End-butt	57*b*	109
Glue blocks	57*d*, 66*b*	109, 115
Scarf	57*c*	109

	Paragraph	Page
Side-grain	57a	109
Loose	66b	115
Splined Miter	17d	64

KILN DRYING. (*See* Lumber, seasoning.)

KITCHEN FURNITURE, FINISHING SCHEDULE ... 80c ... 124

KNOTS. (*See* Lumber, defects.)

LATHE, WOOD. (*See* Wood lathe.)

LIGNIN	3f	3
LOCATING PINS	22	74
LOCKS ON WOOD FURNITURE, REPAIR	65	114
LOGS	3b	2
LONG STOCK, CUTTING With crosscut saw	16b	62

LUMBER

Cutting	4a	4
Edge or vertical grain	4a(2)	4
Flat-grain	4a(1)	4
Plain-sawed	4a(1), 4b(1)	4, 5
Quarter-sawed	4a(2), 4b(2)	4, 5
Defects	5, 6	5, 6
Checks	5c(2)(f)	6
Compression—		
Failure	6e	7
Wood	6d	7
Grain deviation	6b	6
Insect damage	6i	7
Knots	3d, 6c	2, 7
Molds	6h	7
Shakes	6f	7
Stains	6h	7
Warping	5c, 6a, 38b	6, 85
Wood rot	6h	7
Seasoning	5	5
Air drying	5b	6
Chemical	5d	6
Kiln drying	5b, 5c	6
Piling	5b(2)	6

MACHINERY, WOODWORKING	11-53 incl.	54
Band saw	47, 48	96
Circular saw	12-22 incl.	54
Hollow-chisel mortiser	49, 50	100
Overarm saw	23-25 incl.	75
Single surfacer	41, 42, 43	88
Wood jointer	36-40 incl.	84
Wood lathe	51-53 incl.	103
Wood shaper	44-46 incl.	93

	Paragraph	*Page*
MAPLE, BIRD'S-EYE	3c	2
METAL FURNITURE, REPAIR	81	125
Bracing	81b	125
Refinishing	81c	125
Restoring	81a	125
MITERING With—		
Crosscut saw	17	63
Overarm saw	25	75
Single-arbor variety saw	12d	54
MITER SAW. (*See* Circular saws.)		
MOISTURE IN WOOD	3d, 5a, 56	2, 5, 108
MOLDINGS, CUTTINGS with—		
Circular saws	19	67
Single-arbor variety saws	12d	54
Special heads	18d	67
Wood shaper	44	93
MOLDS. (*See* Lumber, defects.)		
MONOCOTYLEDONS	3a(2)	2
MORTISER, HOLLOW-CHISEL:		
Adjustments, clamps and stops	50a, 50b	100
Cutting recess to receive matching tenon	49	100
Mortise	50c	100
MOSS, Use in padding of upholstered furniture	71	117
MUSLIN, Use in upholstering furniture	71	117
OAK	3e	2
OBLIQUE SAWING	19c	68
ORGANIC COMPOUNDS Used for preservation of wood	7	7
OVERARM SAW	23-35 incl.	75
Use:		
Angle cut-off	27	75
Bevel ripping	29	75
Compound miter cuts	28	75
Crosscutting	24	75
Cutting tenons	33	81
Dadoing	30	75
Mitering	25	75
Plowing	31	80
Rabbeting	32	80
Ripping	26	75
Routing	35	81
Shaping	34	81
PADDING FOR UPHOLSTERED FURNITURE:		
Kinds:		
Cotton batting	71a	117

		Paragraph	Page
	Excelsior	71a	117
	Moss	71a	117
	Tow	71a	117
	Replacement	71b	117
PALMS		3a	2
PARALLEL RULE		19c	68
PARTING TOOLS FOR WOOD LATHE. (*See* Wood lathe, tools.)			
PHENOLIC-RESIN GLUE		59g (2)	110
PICTURE FRAMES		17a	63
PILING LUMBER FOR SEASONING		5b (2)	6
PLANER. (*See* Surfacer, single.)			
PLOWING With overarm saw using dado head		31	80
PRESSURE BAR FOR SINGLE SURFACER. (*See* Surfacer, single.)			
PRESSURE USED IN GLUING. (*See* Gluing.)			
PUSH STICK		14g, 17d (3)	56, 64
QUARTER-SAWED LUMBER		4a (2), 4b (2)	4, 5
RABBETING With—			
Overarm saw, using dado head		32	81
Ripsaw		12b, 15b	54
Wood jointer		36, 39	84, 86
Wood shaper		44	93
RECOVERING UPHOLSTERED FURNITURE. (*See* Upholstered furniture, repair.)			
REFINISHING FURNITURE. (*See* Furniture, refinishing.)			
REPAIRING FURNITURE. (*See* Metal furniture, repair; Upholstered furniture, repair;			
Wood furniture, repair.)			
RIPPING AT AN ANGLE With—			
Overarm saw		27	75
Variety saw		15d	59
RIPPING FENCE. (*See* Fence, ripping.)			
RIPPING GAUGE		48b	99
RIPPING, SIMPLE With—			
Circular saw		13a	54
Overarm saw		23, 26	75
Ripsaw		12b, 15a	54, 58
Single-arbor variety saw		12d, 13c	54, 55
Universal double-arbor saw		12c	54
ROOTS		3b, 3e	2
ROUND-NOSE CHISELS. (*See* Wood lathe, tools.)			
ROUTERS		23, 35	75, 81
ROUTING With overarm saw		35	81
SAFETY MEASURES		14, 43h, 44, 46	56, 92, 93, 95

	Paragraph	Page
SALT, Use:		
Drying lumber	5	5
Preservation of wood	7	6
SAPWOOD	3b(2), 55	2, 108
SAW MARKS, REMOVING With single surfacer	41	88
SAWS. (*See* Machinery, woodworking.)		
SHAKES. (*See* Lumber, defects.)		
SHAPERS MOUNTED ON OVERARM SAW	23	75
SHAPING With—		
Overarm saw	34	81
Wood lathe	51	103
Wood shaper	44, 45a, 45b	93, 94
SHRINKAGE OF WOOD	5a(1), 5a(2)	5
SINGLE SURFACE. (*See* Surfacer, single.)		
SKEW CHISELS FOR WOOD LATHE. (*See* Wood lathe, tools.)		
SOAPSTONE, POWDERED Used to relieve sticking of drawers	66a	114
SOFTWOODS	3a(1), 3a(3), 3d, 3e, 4a, 43c, 55	2, 4, 91, 108
SOLID-HEAD CUTTER. (*See* Wood shaper, parts.)		
SOLID-WOOD SURFACES, repair	63	114
SPEAR-POINT CHISELS. (*See* Wood lathe, tools.)		
SPECIAL GROOVING AND MOLDING HEADS	18d	67
SPLINED MITER JOINTS	17d	64
SPRINGS, REPAIR. (*See* Upholstered furniture.)		
SQUARE-NOSE CHISELS. (*See* Wood lathe, tools.)		
SQUARING ENDS With crosscut saw	16a	60
STAINS ON LUMBER. (*See* Lumber, defects.)		
STANDARDS FOR REPAIR WORK	2, 75	1, 120
STARCH GLUE. (*See* glue.)		
SURFACER, SINGLE (or planer)	41, 42, 43	88
Operation	43	90
Board lengths	43e	91
Carrier, using	43g	91
Cut, amount	43c	91
Cut, direction	43d	91
Feeding	43b, 43f	90, 91
Safety measures	43h	92
Scale adjustments	43a	90
Parts:		
Bed	42b	88
Chip breaker	42f	90
Cutter head	42e	90
Driving mechanism	42h	90
Feed rolls	42d	90
Grinding attachments	42i	90

	Paragraph	Page
Pressure bar	42g	90
Raising and lowering device	42c	90
Supporting frame	42a	88

SURFACE PREPARATION IN REFINISHING FURNITURE. (*See* Furniture refinishing.)

SURFACING With—
| Single surfacer or planer | 41 | 88 |
| Wood jointer | 36, 37 | 84 |

| SYCAMORE | 3e | 2 |

TABLE, BAND-SAW. (*See* Band saw.)

TAILSTOCK. (*See* Wood lathe, tools.)

TAPERING With—
Ripsaw, using templates	20b	70
Single surfacer	43g(1)	91
Wood jointer	36	84

TENONS, CUTTING With—
Circular saws	21	70
Overarm saws	33	81
Tenoner	21	70

TENONS, Cheek cuts With—
Combination saw	21c	72
Dado head	21d	73
Ripsaw	21c	72
Shoulder cuts—		
Angled	18b	67
Square	21a	70

| THIN STOCK, PROCESSING With surface carrier | 43g(2) | 91 |

TOOL REST. (*See* Wood lathe, tools.)

| TOW, Use in padding of upholstered furniture | 71 | **117** |

| TREES | 3 | **2** |

TUPELO GUM. (*See* Gum, special drying.)

UPHOLSTERED FURNITURE, REPAIR	69–74 incl.	117
Covering	70	117
Frame	72	117
Padding	71	**117**
Webbing	73, 71b	117

| UREA, Used for drying lumber | 5b, 5d | 6 |

UREA-RESIN GLUE. (*See* glue.)

VEGETABLE PROTEIN GLUE. (*See* glue.)

| VENEERED SURFACES, REPAIR | 64 | **114** |

| WARPED TOPS WOOD FURNITURE, REPAIR | 63d | 114 |

WARPED STOCK, PROCESSING. (*See* Wood jointer, edging.) (*See also* wood jointer, surfacing.)

	Paragraph	Page

WARPING. (*See* Lumber, defects.)

WEBBING, REPAIR. (*See* Upholstered furniture.)

WEDGES, CUTTING With—
Combination saw, using template...20a 69
Ripsaw, using template...20a 69

WIDE STOCK, CUTTING With crosscut saw...16c 62

WOOD:
Composition 3b 2
Heartwood ....3b(2) 2
Preservation ....3b(2), 7 2, 7
Sapwood ....3b(2) 2
Standards for repair work...2a(1), 75 1, 120
Strength 3d 2

WOOD FENCE 48b 99

WOOD FURNITURE, REPAIR...62, 63, 64, 65, 66, 67, 68 114, 115
Drawers 66 114
Fastenings and attachments...67 115
Locks and catches...65 114
Miscellaneous 68 115
Solid surfaces 63 114
Veneered (*See also* furniture refinishing)...64 114

WOOD JOINTER 36, 37, 38, 39, 40 84, 85, 86

Use:
Beveling 40 86
Chamfering 40 86
Edging 38 85
Rabbeting 39 86
Surfacing 37 84
Tapering 36 84

WOOD LATHE 51, 52, 53 103, 104

Parts:
Headstock 51a 103
Tool rest 51b 103
Tailstock 51c 103

Tools:
Calipers and dividers...52g 104
Gouges 52a 104
Parting tools 52c 104
Round-nose chisels 52d 104
Skew chisels 52b 104
Spear-point chisels 52f 104
Square-nose chisels 52e 104

Turning procedure:
Faceplate turning 53b 106
Shaping wood 51 103
Stock held between centers...53a 104

WOOD ROT. (*See* Lumber, defects.)

138

	Paragraph	*Page*
WOOD SHAPER	44, 45, 46	93, 94, 95
Operation:		
Beading	44, 45c	93, 95
Fluting	44, 45c	93, 95
Grooving	44	93
Moldings	44	93
Rabbeting	44	93
Shaping stock:		
With straight edges and external angles	45a	94
With curved edges and internal angles	45b	94
Parts:		
Solid-head cutter	44a	93
Flat-knife cutter	44b	93
Saftey measures	46	95
WOOD SPRING	46	95
WOODWORKING MACHINERY. (*See* Machinery, woodworking.)		
YUCCAS	3a(2)	2
ZINC CHLORIDE, Use for preservation of wood	7	7